FTCE Biology 6-12
Teacher Certification Exam

By: Sharon A. Wynne, M.S.

XAMonline, Inc.
Boston

XAMonline, Inc.
21 Orient Avenue
Melrose, MA 02176
Toll Free 1-800-301-4647
Email: info@xamonline.com
Web www.xamonline.com

Library of Congress Cataloging-in-Publication Data
Wynne, Sharon A.

    FTCE Biology 6-12: Teacher Certification / Sharon A. Wynne.
    ISBN 978-1-64239-002-5

1. Biology 6-12   2. Study Guides.   3. FTCE
4. Teachers' Certification & Licensure.   5. Careers

Disclaimer:

The opinions expressed in this publication are the sole works of XAMonline and were created independently from the National Education Association, Educational Testing Service, or any State Department of Education, National Evaluation Systems or other testing affiliates.

Between the time of publication and printing, state specific standards as well as testing formats and website information may change that is not included in part or in whole within this product. XAMonline developed the sample test questions and the questions reflect similar content as on real tests; however, they are not former tests. XAMonline assembles content that aligns with state standards but makes no claims nor guarantees teacher candidates a passing score. Numerical scores are determined by testing companies such as NES or ETS and then are compared with individual state standards. A passing score varies from state to state.

Printed in the United States of America

FTCE Biology 6-12
ISBN: 978-1-64239-002-5

Table of Contents

Competency 3.0   Knowledge of the chemical processes of living things

Competency   6.0   Knowledge of the structural and functional diversity of viruses and prokaryotic organisms

Competency 9.0 Knowledge of ecological principles and processes

Competency 10.0 Knowledge of evolutionary mechanisms

Great Study and Testing Tips!

What to study in order to prepare for the subject assessments is the focus of this study guide but equally important is *how* you study.

You can increase your chances of truly mastering the information by taking some simple, but effective steps.

Study Tips:

1. Some foods aid the learning process. Foods such as milk, nuts, seeds, rice, and oats help your study efforts by releasing natural memory enhancers called CCKs (*cholecystokinin*) composed of *tryptophan*, *choline*, and *phenylalanine*. All of these chemicals enhance the neurotransmitters associated with memory. Before studying, try a light, protein-rich meal of eggs, turkey, and fish. All of these foods release the memory enhancing chemicals. The better the connections, the more you comprehend.

Likewise, before you take a test, stick to a light snack of energy boosting and relaxing foods. A glass of milk, a piece of fruit, or some peanuts all release various memory-boosting chemicals and help you to relax and focus on the subject at hand.

2. Learn to take great notes. A by-product of our modern culture is that we have grown accustomed to getting our information in short doses (i.e. TV news sound bites or USA Today style newspaper articles.)

Consequently, we've subconsciously trained ourselves to assimilate information better in neat little packages. If your notes are scrawled all over the paper, it fragments the flow of the information. Strive for clarity. Newspapers use a standard format to achieve clarity. Your notes can be much clearer through use of proper formatting. A very effective format is called the *"Cornell Method."*

Take a sheet of loose-leaf lined notebook paper and draw a line all the way down the paper about 1-2" from the left-hand edge.

Draw another line across the width of the paper about 1-2" up from the bottom. Repeat this process on the reverse side of the page.

Look at the highly effective result. You have ample room for notes, a left hand margin for special emphasis items or inserting supplementary data from the textbook, a large area at the bottom for a brief summary, and a little rectangular space for just about anything you want.

3. **Get the concept then the details.** Too often we focus on the details and don't gather an understanding of the concept. However, if you simply memorize only dates, places, or names, you may well miss the whole point of the subject.

A key way to understand things is to put them in your own words. If you are working from a textbook, automatically summarize each paragraph in your mind. If you are outlining text, don't simply copy the author's words.

Rephrase them in your own words. You remember your own thoughts and words much better than someone else's, and subconsciously tend to associate the important details to the core concepts.

4. **Ask Why?** Pull apart written material paragraph by paragraph and don't forget the captions under the illustrations.

Example: If the heading is "Stream Erosion", flip it around to read "Why do streams erode?" Then answer the questions.

If you train your mind to think in a series of questions and answers, not only will you learn more, but it also helps to lessen the test anxiety because you are used to answering questions.

5. **Read for reinforcement and future needs.** Even if you only have 10 minutes, put your notes or a book in your hand. Your mind is similar to a computer; you have to input data in order to have it processed. *By reading, you are creating the neural connections for future retrieval.* The more times you read something, the more you reinforce the learning of ideas.

Even if you don't fully understand something on the first pass, *your mind stores much of the material for later recall.*

6. **Relax to learn so go into exile.** Our bodies respond to an inner clock called biorhythms. Burning the midnight oil works well for some people, but not everyone.

If possible, set aside a particular place to study that is free of distractions. Shut off the television, cell phone, pager and exile your friends and family during your study period.

If you really are bothered by silence, try background music. Light classical music at a low volume has been shown to aid in concentration over other types. Music that evokes pleasant emotions without lyrics are highly suggested. Try just about anything by Mozart. It relaxes you.

7. <u>Use arrows not highlighters</u>. At best, it's difficult to read a page full of yellow, pink, blue, and green streaks. Try staring at a neon sign for a while and you'll soon see that the horde of colors obscure the message.

A quick note, a brief dash of color, an underline, and an arrow pointing to a particular passage is much clearer than a horde of highlighted words.

8. <u>Budget your study time</u>. Although you shouldn't ignore any of the material, *allocate your available study time in the same ratio that topics may appear on the test.*

Testing Tips:

1. **Get smart, play dumb. Don't read anything into the question.** Don't make an assumption that the test writer is looking for something else than what is asked. Stick to the question as written and don't read extra things into it.

2. **Read the question and all the choices _twice_ before answering the question.** You may miss something by not carefully reading, and then re-reading both the question and the answers.

If you really don't have a clue as to the right answer, leave it blank on the first time through. Go on to the other questions, as they may provide a clue as to how to answer the skipped questions.

If later on, you still can't answer the skipped ones . . . **_Guess._** The only penalty for guessing is that you _might_ get it wrong. Only one thing is certain; if you don't put anything down, you will get it wrong!

3. **Turn the question into a statement.** Look at the way the questions are worded. The syntax of the question usually provides a clue. Does it seem more familiar as a statement rather than as a question? Does it sound strange?

By turning a question into a statement, you may be able to spot if an answer sounds right, and it may also trigger memories of material you have read.

4. **Look for hidden clues.** It's actually very difficult to compose multiple-foil (choice) questions without giving away part of the answer in the options presented.

In most multiple-choice questions you can often readily eliminate one or two of the potential answers. This leaves you with only two real possibilities and automatically your odds go to Fifty-Fifty for very little work.

5. **Trust your instincts.** For every fact that you have read, you subconsciously retain something of that knowledge. On questions that you aren't really certain about, go with your basic instincts. **Your first impression on how to answer a question is usually correct.**

6. **Mark your answers directly on the test booklet.** Don't bother trying to fill in the optical scan sheet on the first pass through the test.

Just be very careful not to miss-mark your answers when you eventually transcribe them to the scan sheet.

7. **Watch the clock!** You have a set amount of time to answer the questions. Don't get bogged down trying to answer a single question at the expense of 10 questions you can more readily answer.

Competency 1.0   Knowledge of the investigative processes of science

Skill 1.1   Demonstrate knowledge of the proper use and care of the light microscope.

Light microscopes are commonly used in high school laboratory observations and experiments. Total magnification is determined by multiplying the ocular (usually 10X) and the objective (usually 10X on low, 40X on high) lenses. Several procedures should be followed to properly care for this equipment.

-Clean all lenses with lens paper only.
-Carry microscopes with two hands; one on the arm and one on the base.
-Always begin focusing on low power, then switch to high power.
-Store microscopes with the low power objective down.
-Always use a coverslip when viewing wet mount slides.
-While looking from the side, not through the ocular, bring the objective down to its lowest position then look through the ocular and focus moving the objective up to avoid -breaking the slide or scratching the lens.

Skill 1.2   Recognize and distinguish between the types of microscopy and uses.

Anton van Leeuwenhoek is known as the father of microscopy. In the 1650s, Leeuwenhoek began making tiny lenses which gave magnifications up to 300x. He was the first to see and describe bacteria, yeast plants, and the microscopic life found in water. Over the years, light microscopes have advanced to produce greater clarity and magnification. The transmission electron microscope (TEM) was developed in the 1950s. Instead of light, a beam of electrons passes through the specimen. The scanning electron microscope (SEM) is similar, but the electrons bounce off the surface of the sample rather than passing through it, allowing a view of the surface topography of cells. Scanning electron microscopes have a resolution about one thousand times greater than light microscopes. The disadvantage of the TEM and SEM is that the chemical and physical methods used to prepare the sample result in the death of the specimen.

Skill 1.3   Identify common laboratory techniques (e.g., dissection; preserving, staining, and mounting microscope specimens; preparing laboratory solutions).

Some of the most common laboratory techniques are: dissections, preserving, staining and mounting microscopic specimens, and preparing laboratory solutions.

1. Dissections

Animals that are not obtained from recognized sources should not be used. Decaying animals or those of unknown origin may harbor pathogens and/or parasites. Specimens should be rinsed before handling. Latex gloves are desirable. If gloves are not available, students with sores or scratches should be excused from the activity. Formaldehyde is a carcinogenic and should be avoided or disposed of according to district regulations. Students objecting to dissections for moral reasons should be given an alternative assignment.

Live specimens - No dissections may be performed on living mammalian vertebrates or birds. Lower order life and invertebrates may be used. Biological experiments may be done with all animals except mammalian vertebrates or birds. No physiological harm may result to the animal. All animals housed and cared for in the school must be handled in a safe and humane manner. Animals are not to remain on school premises during extended vacations unless adequate care is provided. Many state laws state that any instructor who intentionally refuses to comply with the laws may be suspended or dismissed. Interactive dissections are available online or from software companies for those students who object to performing dissections. There should be no penalty for those students who refuse to physically perform a dissection.

2. Staining

Specimens have to be stained because they are mostly transparent (except plant cells which are green) under the microscope and are difficult to be seen under microscope against a white background. The stains add color to the picture, making the image much easier to see. The stains actually work by fixing themselves to various structures on or in the cell. The exact structure determines the staining process used.

The variety of stains available are numerous, and are a vital tool to determine what the cellular components are made of. Starch, protein and even nucleic acids can be brought out using special stains.

Some common stains used in the laboratories are: methylene blue, chlorazol black, lignin pink, gentian violet, etc.

3. Mounting of specimens

In order to observe microscopic specimens or minute parts, mounting them on a microscope slide is essential. There are two different ways of mounting. One kind of procedure is adapted for keeping mounted slides for a long time to be used again.

The second type of procedure is for temporary slides. We will discuss temporary mounting since 12th Grade students are mostly concerned with the temporary mounting. Their work does not require permanent mounting.

Water is a very common mounting medium in high school laboratories since it is cheap and best suited for temporary mounting. One problem with water mounting is that water evaporates.

Wet mount slides should be made by placing a drop of water on the specimen and then putting a glass coverslip on top of the drop of water. Dropping the coverslip at a forty-five degree angle will help in avoiding air bubbles.

Glycerin is also used for mounting. One advantage with glycerin is that it is non-toxic and is stable for years. It provides good contrast to the specimens under microscopic examination. The only problem with glycerin as a medium is that it supports mold formation.

4. Preparation of laboratory solutions

This is a critical skill needed for any experimental success. The procedure for making solutions must be followed to get maximum accuracy.

   i)   weigh out the required amount of each solute
   ii) dissolve the solute in less than the total desired volume (about 75%)
   iii) add enough solvent to get the desired volume

1. Weight/volume:

Usually expressed as mg/ml for small amounts of chemicals and other specialized biological solutions. e.g. 100 mg/ml ampicillin = 100 mg. of ampicillin dissolved in 1 ml of water.

2. Molarity: moles of solute dissolved/ liter of solution

   Mole = 6.02 times 10^{23} atoms = Avogadro's number
   Mole = gram formula weight (FW) or gram molecular weight (MW)
    * These values are usually found on the labels or in Periodic Table.
      e.g. Na_2SO_4

2 sodium atoms - 2 times 22.99g =   45.98 g
1 sulfur atom   - 1 times 32.06g =   32.06 g
4 oxygen atoms – 4 times16.00g =   64.00 g
   Total   =   142.04g

1M = 1 mole/liter, 1 mM = 1 millimole/liter, 1 uM = 1 umole/liter

* How much sodium is needed to make 1L of 1M solution?
Formula weight of sodium sulfate = 142.04g
Dissolve 142.04g of sodium sulfate in about 750mL of water, dissolve sodium sulfate thoroughly and make up the volume to 1 liter (L)

Skill 1.4   Identify proper field techniques (e.g., site selection, field procedures, sampling, capture/recapture, transects, collecting techniques, environmental quality assessment).

Some of the field techniques described here are site selection, field procedures, sampling, capture/recapture, transects, collecting techniques, and environmental assessment. We will look at each of these individually.

Site selection:
Site selection in any field experiment is a critical factor. It depends on a number of issues including the type of research that will be conducted in that site, the duration of the investigation, and the accessibility of the site to city/town and transportation to that site. The selection of a site is determined by a group of researchers from the team who study all of the site aspects, make their recommendations, and then site selection may be approved by the research team. If the first choice is found to be unsuitable a second site, more suitable, will be selected. The most important thing is that all the researchers must be reasonably comfortable with the site.

Field procedures:
Field procedures are procedures that are done for a successful sample collection. These are:

1. Preparing for field study: identifying learning objectives, the purpose of field study
2. Site selection: finding out a suitable site for the type of investigation
3. Sample collection: collecting samples
4. Preserving collected specimens

Sampling:
Sampling is collecting pieces/specimens or making instrument data points/observations at determined intervals or areas for the purpose of research/investigation. Sampling includes animal tracking, capturing, plant and animal tagging, plot sampling, specimen collecting, transect sampling, water sampling, etc. The results obtained are used as representative of the whole research area or population.

Capture/recapture:
Capture /recapture are methods very commonly used in ecological studies. This method is also known as mark/capture, capture-mark-recapture, sight-resight and band recovery.

A researcher visits the study area (see site selection) and uses traps to capture organisms alive. Each of these is marked with a unique identifier- a numbered tag/band- and then is released unharmed back into the environment. Next, the researcher returns and captures another sample of organisms. Some of the organisms in the second sample will have been marked during the first visit and are known as recaptures. The unmarked organisms are tagged just like the previous ones.

Population size can be estimated from as few as two visits to the study area, but usually more than two visits are made.

Transect:
A transect is a path along which one records and/or counts occurrences of the phenomenon of study (e.g. animals, for instance by noting each individual animal's distance from the path, species of plants, in the process of estimating population densities in a study area). This action requires an observer to move along a fixed path and to count occurrences along the path and, at the same time, obtain the distance of the object from the path. This results in an estimate of the area covered, an estimate of the way in which detectability drops off from probability 1 to 0 as one moves away from the path. Using these two figures one can arrive at an estimate of the actual density of objects.

Collecting techniques:
The objective of capturing an animal is to identify it. It is much easier this way since the animal is in our possession. There are 6 steps involved in this seemingly simple procedure:

1. Catching – it is very difficult to catch insects like butterflies and dragon flies since they are very quick and the person trying to catch them must be quicker. The easy way is to use a net. This is safe and quick and eliminates running around with the insects.

2. Enveloping – Place the dragon/butter fly wings folded in a glassine (stamp/coin collection bag), label it, and number it. This way, when this bag is opened at a later date, the information needed is on the envelope.

3. Acetoning – After leaving the specimens for awhile (in the mean time, they will empty their intestines), one sacrifices the specimen by immersing it in acetone briefly. Straighten the insect and then return it to the acetone and leave it there for 16-24 hours. Acetone extracts water and fat from the specimens and they dry much better, though some specimens may become discolored.

4. Removing – Remove from acetone and allow the specimens to dry in a spot away from people, since acetone fumes are not good for our health.

5. Labeling – All the pertinent information must be attached to the specimens so that information is much clearer at a later date when someone else is studying them.

6. Storing – The preserved specimens must be stored in a box protected from humidity and pests.

Environmental quality assessment:
Environmental assessment is to study, collect information, and analyze it using scientific principles and evaluate the quality and conditions of the type of environment under study – e.g., marine, coastal, lake, etc. This is absolutely essential to determine the hazards that are causing pollution and their effect on human and other life forms. The environmental quality assessment is done in a site or multi site, for a short period of time or lasting for a number of years. Whatever may be the size of the test site or the duration of the investigation, the basic aim is to determine the quality of that particular environment. Some environmental projects study water quality, air quality, sediment and soil assessment, ground water assessment, oil spills and their effect on marine life, etc.

 Skill 1.5 Identify the uses of PCR, chromatography, spectrophotometry, centrifugation, and electrophoresis.

Chromatography refers to a set of techniques that are used to separate substances based on their different properties such as size or charge. Paper chromatography uses the principles of capillarity to separate substances such as plant pigments. Molecules of a larger size will move more slowly up the paper, whereas smaller molecules will move more quickly producing lines of pigment.

An **indicator** is any substance used to assist in the classification of another substance. An example of an indicator is litmus paper. Litmus paper is a way to measure whether a substance is acidic or basic. Blue litmus turns pink when an acid is placed on it and pink litmus turns blue when a base is placed on it. pH paper is a more accurate measure of pH, with the paper turning different colors depending on the pH value.

Spectrophotometry measures the percent of light at different wavelengths absorbed and transmitted by a pigment solution.

Centrifugation involves spinning substances at a high speed. The more dense part of a solution will settle to the bottom of the test tube, where the lighter material will stay on top. Centrifugation is used to separate blood into blood cells and plasma, with the heavier blood cells settling to the bottom.

Electrophoresis uses electrical charges of molecules to separate them according to their size. The molecules, such as DNA or proteins are pulled through a gel toward either the positive end of the gel box (if the material has a negative charge) or the negative end of the gel box (if the material has a positive charge). DNA is negatively charged and moves towards the positive charge.

One of the most widely used genetic engineering techniques is **polymerase chain reaction (PCR)**. PCR is a technique in which a piece of DNA can be amplified into billions of copies within a few hours. This process requires primer to specify the segment to be copied, and an enzyme (usually taq polymerase) to amplify the DNA.

PCR has allowed scientists to perform several procedures on the smallest amount of DNA.

Skill 1.6   Identify terms in a formula (e.g., chemical, ecological, physical) and assess the relationships among the terms.

A formula is a shorthand way of showing what is in a compound by using symbols and subscripts. The letter symbols let us know the elements that are involved and the number subscripts tell us how many atoms of each element are involved. No subscript is used if there is only one atom involved.
For example – CH_4 – This compound is methane gas and it has one carbon atom and 4 hydrogen atoms.

1. Chemical formulas:
Aerobic respiration: Let us look at this example. Our tissues need energy for growth, repair, movement, excretion, and so on. This energy is obtained from glucose supplied to the tissues by our blood. In aerobic respiration, glucose is broken down in the presence of oxygen. The atoms are rearranged into the smaller molecules of carbon dioxide and water, and energy is released, which is used for our metabolic processes. The energy released was previously the chemical bond energy holding the atoms of glucose together.
The above reaction can be written in the form of a word reaction:

Glucose + Oxygen = Carbon Dioxide + Water + Energy

By using chemical symbols and subscripts we can rewrite the above word equation into a proper chemical equation:

$$C_6H_{12}O_6 + 6O_2 = 6CO_2 + 6H_2O + Energy$$

The compounds on the left side of the equation are called reactants and the compounds on the right side of the reaction are called products. Because matter (atoms) are not being created or destroyed in a chemical reaction, the numbers of atoms on either side of the equation are equal. The reactants in the above equation have to combine in a fixed proportion for a chemical reaction to take place.

2. Ecological formulas:
A number of formulas are used in ecological research. We will use one of the most widely used formulas in ecological investigations for an example. It is called the Lincoln-Petersen method of analysis, and is used in population estimations.

$N = (n_1 n_2)/m$

Where

N = Estimate of total population size

n_1 = Total number of animals captured in the first visit

n_2 = Total number of animals captured on the second visit

m = Number of animals captured on the first visit that were then recaptured on the second visit

3. Physical formulas:

There are a number of formulas that are used in Physical Science and Physics. Let us look at a very simple one –

D = m/v

Where

D = density g/cm

m = mass in grams

v = volume in cm

The above formula is used for calculating the density of an object.

It is absolutely important to write the appropriate units e.g., g (gram), cm (centimeter) etc.

The second example is the formula for calculating the momentum of an object.

M = mass (kg) times velocity (meters/second)

M = mv

The units of momentum are kg (m/s)

The above are only two of the many formulas that are used in a Physical science classroom.

Skill 1.7   Identify the units in the metric system and convert between dimensional units for one-, two-, and three-dimensional objects.

Science uses the **metric system**; as it is accepted worldwide and allows easier comparison among experiments done by scientists around the world.

The meter is the basic metric unit of length. One meter is 1.1 yards. The liter is the basic metric unit of volume. 1 gallon is 3.846 liters. The gram is the basic metric unit of mass. 1000 grams is 2.2 pounds.

The following prefixes are used to describe the multiples of the basic metric units.

Prefix	Multiplying factor	Prefix	Multiplying factor
deca-	10X the base unit	deci-	1/10 the base unit
hecto-	100X	centi-	1/100
kilo-	1,000X	milli-	1/1,000
mega-	1,000,000X	micro-	1/1,000,000
giga-	1,000,000,000X	nano-	1/1,000,000,000
tera-	1,000,000,000,000X	pico-	1/1,000,000,000,000

Skill 1.8   Identify assumptions, observations, hypotheses, conclusions, and theories and differentiate between them.

Science may be defined as a body of knowledge that is systematically derived from study, observations, and experimentation. Its goal is to identify and establish principles and theories that may be applied to solve problems. Pseudoscience, on the other hand, is a belief that is not warranted. There is no scientific methodology or application. Some of the more classic examples of pseudoscience include phrenology, astrology, or any topic that is explained by hearsay.

Assumptions are conditions or properties assumed to be true or consistent. In any experimental situation, assumptions are made. When designing an experiment, assumptions being made should be identified and judged as reasonable. For example, if studying tomato plants, a reasonable assumption is that tomatoes of the same variety are genetically similar and as a group may show certain tendencies. An unreasonable assumption might be that by testing only one tomato plant we can conclude how all other tomato plants of that variety would respond in similar conditions.

Observations are made either directly through human senses or indirectly with the help of tools and technology such as oxygen sensors, microscopes, cameras, etc. Observations may yield either quantitative or qualitative data. All observations are systematically recorded.

Conclusions are drawn based on the data gathered in an experiment and are related to the hypothesis. Scientific reporting of conclusions includes an analysis of possible sources of error and suggestions for further experimentation to further test the conclusions.

Scientific theory and experimentation must be repeatable. It is also possible to be disproved and is capable of change. Science depends on communication, agreement, and disagreement among scientists. It is composed of theories, laws, and hypotheses.

theory - the formation of principles or relationships which have been verified and accepted. A theory accurately predicts experimental results, and adequately explains all current observations. Therefore, in scientific language a theory is never "just" a theory.

law - an explanation of events that occur with uniformity under the same conditions (laws of nature, law of gravitation).

hypothesis - an idea, proposed explanation, or educated guess followed by research which can test it. A   proven hypothesis is one of the building blocks of a theory.One experiment can support a hypothesis; many experiments and observations are used to support a theory.

Skill 1.9   Evaluate, interpret, and predict from data sets, including graphical data.

Graphing is an important skill to visually display collected data for analysis. The two types of graphs most commonly used are the **line graph** and the **bar graph** (histogram). Line graphs are set up to show two variables represented by one point on the graph. The X-axis is the horizontal axis and represents the independent variable. Independent variables are those that would be present independently of the experiment. A common example of an independent variable is time. Time proceeds regardless of anything else going on. Other independent variables are experimental factors manipulated by the scientist to test a hypothesis, such as nutrient quantity or drug dosage. The Y-axis is the vertical axis and represents the dependent variable. Dependent variables, such as growth of a plant or an immune response, are ones that may vary based on experimental factors. Graphs should be calibrated at equal intervals. If one space represents one day, the next space may not represent ten days. A "best fit" line is drawn to join the points and may not include all the points in the data. Axes must always be labeled. A good title will describe both the dependent and the independent variable. Bar graphs are set up similarly in regards to axes, but points are not plotted. Instead, the dependent variable determines the height of a bar whose position on the X-axis indicates the independent variable. . Each bar is a separate item of data and is not joined by a continuous line.

The type of graphic representation used to display observations depends on the data that is collected. **Line graphs** are used to compare different sets of related data or to predict data that has yet been measured. An example of a line graph would be comparing the rate of activity of different enzymes at varying temperatures. A **bar graph** or **histogram** is used to compare different items and make comparisons based on this data. An example of a bar graph would be comparing the ages of children in a classroom. A **pie chart** is useful when organizing data as part of a whole. A good use for a pie chart would be displaying the percent of time students spend on various after school activities.

Analyzing graphs is a useful method for determining the mathematical relationship between the dependent and independent variables of an experiment. The usefulness of the method lies in the fact that the variables represent on the axes of a straight-line graph are represented by the expression, $y = m*x + b$, where m=the slope of the line, b=the y intercept of the line. This equation works only if the data fit a straight-line graph.

Thus, once the data set has been collected, and modified, and plotted to achieve a straight-line graph, the mathematical equation can be derived.

Skill 1.10   Differentiate the characteristics of scientific research from other areas of learning.

Scientific research serves two purposes –

1. To investigate and acquire knowledge which is theoretical and
2. To do research which is of practical value.

Science is in a unique position to be able to serve humanity. Scientific research comes from inquiry. An inquiring mind is thirsty, trying to find answers. The two most important questions – why and how are the starting points. A person who is inquisitive asks questions and wants to find out answers.

The characteristics of scientific research are methodical and are very different from learning other areas like mathematics and history, to name a few.

Scientific research uses scientific method to answer the questions. Those who research follow the scientific method, which consists of a series of steps designed to solve a problem or find answer to their problem.

The aim of the scientific method is to eliminate bias / prejudice from the scientist/researcher. As human beings, we are influenced by our bias/prejudice and this method helps to eliminate that. If all the steps of the scientific method are followed as outlined, there is the maximum elimination of bias.

The scientific method is made up of the following steps –

1. Stating the problem clearly and precisely
2. Gathering information/research
3. Hypothesis (an educated guess)
4. Experimental design
5. Analysis of the results
6. Conclusion

Scientific research is clearly different from the learning of other areas. Science demands evidence. Science requires experimenting to prove one's ideas or propositions. Science does not answer all our questions. Science doesn't say anything about the cultural, moral and religious beliefs of the individuals. It is up to us to use the information science provides and make our own decisions according to our beliefs and norms.

The learning of science is based on inquiry. Inquiry is absolutely important to researchers because it is the starting point of learning. Researchers have a responsibility to society in making them literates in science (to understand everyday science problems like the greenhouse effect, pollution, energy crisis, unconventional energy etc.,) and to pass on the information, which are experimental findings to the society for its benefit and use.

Lastly researchers are bound by guidelines based on ethical, moral and cultural issues.

Skill 1.11 Distinguish between accuracy and precision, and between systematic error and random error.

Accuracy and precision

Accuracy is the degree of conformity of a measured, calculated quantity to its actual (true) value. Precision also called reproducibility or repeatability and is the degree to which further measurements or calculations will show the same or similar results. Accuracy is the degree of veracity while precision is the degree of reproducibility.

The best analogy to explain accuracy and precision is the target comparison. Repeated measurements are compared to arrows that are fired at a target. Accuracy describes the closeness of arrows to the bull's eye at the target center. Arrows that strike closer to the bull's eye are considered more accurate.

Systematic and random error

All experimental uncertainty is due to either random errors or systematic errors. Random errors are statistical fluctuations in the measured data due to the precision limitations of the measurement device. Random errors usually result from the experimenter's inability to take the same measurement in exactly the same way to get exactly the same number. Systematic errors, by contrast, are reproducible inaccuracies that are consistently in the same direction. Systematic errors are often due to a problem, which persists throughout the entire experiment. Systematic and random errors refer to problems associated with making measurements. Random errors, since they occur in both directions randomly, likely cancel out especially when sampling sizes are large enough. Systematic errors, however, can lead to inaccurate results. Mistakes made in the calculations or in reading the instrument are not considered in error analysis.

Skill 1.12 Characterize variables and the outcomes for appropriate experimental designs.

The procedure used to obtain data is important to the outcome. Experiments consist of **controls** and **variables**. A control is the experiment run under normal conditions. The variable includes a factor that is changed. In biology, the variable may be light, temperature, pH, time, etc. The differences in tested variables may be used to make a prediction or form a hypothesis. Only one variable should be tested at a time. One would not alter both the temperature and pH of the experimental subject.

An **independent variable** is one that is changed or manipulated by the researcher. This could be the amount of light given to a plant or the temperature at which bacteria is grown. The **dependent variable** is that which is influenced by the independent variable.

Skill 1.13   Recognize that the validity and reliability of scientific knowledge is based on reproducibility of results and statistical significance of results, and is limited by the state of current technology and possible bias.

The validity of the scientific knowledge is based in the reproducibility of results and statistical significance. Results must be reproducible regardless of whether it is of practical use or of theoretical application.

Investigations yield data, which are then analyzed. The experiments must be repeated a few times, at least twice, to get reliable data. Reliable data are data that are reproducible and from which theories are formulated. Anybody who does the same experiment anywhere in the world must get same type result. It must follow the pattern, although not the exact same figures. Patterns are what researchers look for in their investigations. Hence, reproducibility is very important in scientific investigations.

The data obtained in investigations are subjected to statistical analysis. In order for a conclusion to be drawn either supporting or refuting a hypothesis, the data must be statistically significant. With the advancement of technology, scientists are able to conduct highly sophisticated experiments. But sometimes, the available technology is not sophisticated enough for the type of research that is planned. Scientists are constantly looking for higher levels of technology for use in experimentation. Hence, we can say that research is being limited by the technology.

Scientists, being human, are subject to bias, even if subconscious and unintentional. A personal agenda to find a cure for cancer might lead to systematic errors in the collection of data, for example. Peer review and repetition of experimentation are two ways that science guards against possible bias.

Skill 1.14   Identify the development of biological knowledge through important historical events, individuals (e.g., Robert Hooke, Mattias Schleiden, Francis Jacob, Jacques Monod), and experimental evidence.

The history of biology traces mans' understanding of the living world from the earliest recorded history to modern times. Though the concept of biology as a field of science arose only in the 19th century, the origin of biological sciences could be traced back to ancient Greeks (Galen and Aristotle).

During the Renaissance and Age of Discovery, renewed interest in the rapidly increasing number of known organisms generated lot of interest in biology.

Andreas Vesalius (1514-1564) was a Belgian anatomist and physician whose dissections of human body and descriptions of his findings helped to correct the misconceptions of science. The books Vesalius wrote on anatomy were the most accurate and comprehensive anatomical texts to date.

Anton van Leeuwenhoek is known as the father of microscopy. In the 1650s, Leeuwenhoek began making tiny lenses that gave magnifications up to 300x. He was the first to see and describe bacteria, yeast plants, and the microscopic life found in water. Over the years, light microscopes have advanced to produce greater clarity and magnification.

Robert Hooke (1635-1703) was a renowned inventor, a natural philosopher, astronomer, experimenter and a cell biologist. He deserves more recognition than he had, but he is remembered mainly for Hooke's law, an equation describing elasticity that is still used today. He was the type of scientist that was then called a "virtuoso"- able to contribute findings of major importance in any field of science. Hooke published *Micrographia* in 1665. Hooke devised the compound microscope and illumination system, one of the best such microscopes of his time, and used it in his demonstrations at the Royal Society's meetings. With it he observed organisms as diverse as insects, sponges, bryozoans, foraminifera, and bird feathers. *Micrographia* is an accurate and detailed record of his observations, illustrated with magnificent drawings.

Carl Von Linnaeus (1707-1778), a Swedish botanist, physician and zoologist is well known for his contributions in ecology and taxonomy. Linnaeus is famous for his binomial system of nomenclature in which each living organism has two names, a genus and a species name. He is considered as the father of modern ecology and taxonomy.

In the late 1800s, Pasteur discovered the role of microorganisms in the cause of disease, which led him to develop pasteurization and the rabies vaccine. Koch took these observations one step further by formulating that specific diseases were caused by specific pathogens. **Koch's postulates** are still used as guidelines in the field of microbiology: the same pathogen must be found in every diseased person, the pathogen must be isolated and grown in culture, the disease is induced in experimental animals from the culture, and the same pathogen must be isolated from the experimental animal.

Mattias Schleiden, a German botanist is famous for his cell theory. He observed plant cells microscopically and concluded that cell is the common structural unit of plants. He proposed the cell theory along with Schwann, a zoologist, who observed cells in animals.

In the 18th century, many fields of science like botany, zoology and geology began to evolve as scientific disciplines in the modern sense.

In the 20th century, the rediscovery of Mendel's work led to the rapid development of genetics by Thomas Hunt Morgan and his students.

DNA structure was another key event in biological study. In the 1950s, James Watson and Francis Crick discovered the structure of a DNA molecule as that of a double helix. This structure made it possible to explain DNA's ability to replicate and to control the synthesis of proteins.
Francois Jacob and Jacques Monod contributed greatly to the field of lysogeny and bacterial reproduction by conjugation and both of them won the Nobel Prize for their contributions.

Following the cracking of the genetic code, biology has largely split between 1) organismal biology, consisting of ecology, ethology, systematics, paleontology, and evolutionary biology, developmental biology, and other disciplines that deal with whole organisms or group of organisms; and 2) the disciplines related to molecular biology , including cell biology, biophysics, biochemistry, neuroscience, immunology, and many other overlapping subjects.

The use of animals in biological research has expedited many scientific discoveries. Animal research has allowed scientists to learn more about animal biological systems, including the circulatory and reproductive systems. One significant use of animals is for the testing of drugs, vaccines, and other products (such as perfumes and shampoos) before use or consumption by humans. Along with the pros of animal research, the cons are also very significant. The debate about the ethical treatment of animals has been ongoing since the introduction of animals in research. Many people believe the use of animals in research is cruel and unnecessary. Animal use is federally and locally regulated. The purpose of the Institutional Animal Care and Use Committee (IACUC) is to oversee and evaluate all aspects of an institution's animal care and use program.

Skill 1.15 Differentiate between qualitative and quantitative data in experimental, observational, and modeling methods of research.

. Quantitative data exits in numerical form and can be statistically analyzed. Qualitative data exits in the form of words that describe appearance, behavior, processes, etc. Qualitative data allows for the communication of details that may be harder to communicate with numbers, and might be especially helpful when describing individual cells, organisms, or processes. Quantitative data is more useful when sample sizes are large and measures of central tendency and deviation are desired. It should be noted that much data often thought of as qualitative, such as color, can be quantified, in this example by using a numerical wavelength. Likewise, scientists can decide to quantify other data that is at first collected qualitatively. For example, qualitative descriptions of animal aggression might be divided into five different levels and rated with the numerals

1 – 5. Whether quantitative or qualitative data is used in research depends on the nature of the things being studied, the hypothesis being tested, the practical aspects of reporting data for clarity and completeness, and the desired future use of the data. If is very common that both quantitative and qualitative data are reported in biological research.

Skill 1.16   Recognize the elements of a well-designed and controlled experiment.

Designing an experiment properly is absolutely important because the success of the experiment depends on it. Before designing an experiment, one must identify the elements of an experiment.

1. Control/standard: A control is something the results of an experiment are compared with. Without this, we have no clue of the significance of the data obtained in an experiment.
2. Constants: We need to have many constants for better results. Many factors have to be kept constant in an experiment. The reliability of the data/results depends to a greater extent on the number of constants. So, it is very important to identify all the possible constants in an experiment, which then will be called a well-controlled experiment.
3. Independent variables: These are the variables we have the power to change. It is entirely up to us to choose the independent variables. These factors are going to influence the outcome of the experiment. One has to keep in mind that the number of independent variables must be limited to a maximum of four, otherwise the experiment gets complicated and the data may not be reliable.
4. Dependent variable: This is the factor that will be measured in an experiment. It is called dependent variable since its outcome is dependent on the independent variables.

After the experiment is conducted, it is of utmost importance to repeat the experiment at least twice to obtain reliable data, which then can be analyzed so that conclusions can be drawn. Repetition of the experiment by other scientists in other locations is also important in order to validate results.

Inferring is a very important skill since it interprets the results and facilitates the researcher/scientist to draw logical conclusions.

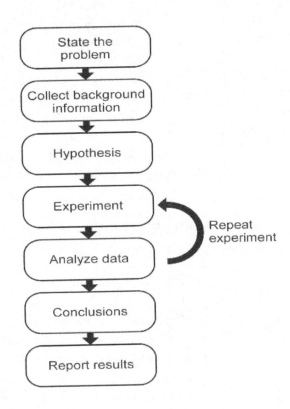

Lastly, there is another important element to the experiment. The conclusions drawn must be communicated orally and in written form for the benefit of furthering knowledge and sharing with the community to enlighten and educate it scientifically. This will help the society in the long run to become scientifically literate.

Skill 1.17 **Identify evidence of the evolutionary nature of science in the face of new observations.**

Science can be understood as a large body of organized knowledge and information that has been gathered as a result of experimentation and observation. It is therefore a group effort made over time by people who seek to slowly improve their understanding of the natural world. Therefore, science has an evolutionary nature; that is, it will change over time as more knowledge is added through continued experimentation and observation. Since experimentation and observation are often assisted with the latest technologies, the accuracy and completeness of gathered information generally improves over time.

When all current observations and experiments support a particular explanation for natural events, that explanation is known as a scientific theory. Therefore, the use of the word "theory" in science is quite different from how it is sometimes used in nonscientific language, when people are sometimes heard saying something is "just" a theory.

An example of scientific knowledge changing over time is the idea that all living cells come from other living cells. Though this is widely accepted today and has been supported through all modern experiments and observations (thereby becoming a part of the Cell Theory), it was not always understood to be true. Without microscopes and constant laboratory observations, past generations believed in spontaneous generation. This is the idea that life can arise from nonliving matter. The evidence for this might have been the lifeless carcass, seen a day later teeming with living maggots. As science advanced and observations improved, the understanding that the living maggots came from living flies' eggs was added to science, and the idea of spontaneous generation was dropped from science.

Skill 1.18   Identify the consistent patterns that govern the occurrence of most natural events.

Natural events within living systems follow certain patterns. For example, all living things reproduce, and this ability to do so is rooted in the unique DNA structure which allows self-replication of the genetic material so that each daughter cell receives a complete set of genes. Just as cells can reproduce, so can each life form, using the mechanisms of cellular reproduction (mitosis and meiosis) to do so.

Another pattern is the flow of energy through living systems. All energy for life originates in the sun; life on earth would be impossible without sunlight. This energy flows from the sun, through chlorophyll, into chemical bonds holding glucose and other organic molecules together, then into the high-energy bonds of ATP. From those bonds, the energy is released once again by cells to do the work of staying alive.

Although energy flow through living systems follows a one-way pattern that requires constant energy input, the pattern of matter flow is different. Matter is constantly recycled in living systems. For example, at the cellular level cells re-use ADP and phosphate groups to make ATP; when the ATP is broken back down to release energy, the ADP and phosphate are used over again. Matter is also recycled on a larger scale, in ecosystems, where nitrogen, carbon, and other nutrients move through individual organisms, into the environment, and back into individual organisms again.

Another pattern of natural events in living systems is fluctuation and homeostasis. Fluctuations within cells, individual organisms, and ecosystems may occur, but if life is to continue these fluctuations need to remain within tolerance levels. The tendency to stay within tolerance levels can be called homeostasis. An example of a fluctuation within an ecosystem might be the decline in population of a top predator species. This would then cause herbivores to increase, which might result in the destruction of certain plant species, which might then cause the overpopulation of other plant species. Such population shifts are often self-correcting over time unless an extreme environmental change or the extinction of a species has occurred.

Another pattern of natural events in living systems is the adaptation of populations to changing environments over time through the process of natural selection, discussed in more depth in Competency 10.

Competency 2.0   **Knowledge of the interaction of the Science, Technology, and Society including Ethical, Legal and Social Issues**

Skill 2.1   **Identify and analyze areas of scientific research that may contribute to ethical, legal, and social conflicts (e.g., reproductive and life-sustaining technologies; genetic basis for behavior, population growth and control; government and business influences on biotechnology).**

With advances in biotechnology come those in society who oppose it. Ethical questions come into play when discussing animal and human research. Does it need to be done? What are the effects on humans and animals? There are no right or wrong answers to these questions. There are governmental agencies in place to regulate the use of humans and animals for research.

Society depends on science, yet it is necessary that the public be scientifically literate and informed in order to prevent potentially unethical procedures from occurring. Especially vulnerable are the areas of genetic research, fertility, and life sustaining medical treatment. It is important for science teachers to stay abreast of current research and to involve students in critical thinking and ethics whenever possible.

Skill 2.2   **Identify principles and uses of cloning, genomics, proteomics, and genetic engineering and analyze possible ethical conflicts.**

Society impacts biological research. The pressure from the majority of society has led to these bans and restrictions on human cloning research. Human cloning has been restricted in the United States and many other countries. The U.S. legislature has banned the use of federal funds for the development of human cloning techniques. Some individual states have banned human cloning regardless of where the funds originate.

Genetic engineering has drastically advanced biotechnology. With these advancements come concerns for safety and ethical questions. Many safety concerns have answered by strict government regulations. The FDA, USDA, EPA, and National Institutes of Health are just a few of the government agencies that regulate pharmaceutical, food, and environmental technology advancements.

Several ethical questions arise when discussing biotechnology. Should embryonic stem cell research be allowed? Is animal testing humane? There are strong arguments for both sides of the issues and there are some government regulations in place to monitor these issues.

Skill 2.3 **Recognize and analyze global environmental challenges that may result from scientific and technological advances and the subsequent resolution of these problems (e.g., CFCs as coolants and ozone depletion; insecticides for protecting crops and pollution events).**

Global environment is posing a huge challenge to governments of all countries. Our technological and scientific advances have drastically affected our environment. Many governments are trying to face this challenge by using technology and also by passing laws to stop some undesirable practices that are polluting our planet.

The governments of some nations (including, until recently, the U.S.) are taking this problem very seriously. They are concerned with multiple things: to allow industrial growth while restoring the environment and ensuring the health of humans. In order to understand this problem, we need to look at some of the contemporary issues like - CFCs as coolants, ozone depletion, insecticides for protecting crops, and pollution events.

1. CFCs as coolants:

Chlorofluorocarbons (CFCs) are a family within the group called haloalkanes (also known as halogenoalkanes). This group of chemical compounds consists of alkanes (such as methane or ethane), with one or more halogens (chlorine or fluorine) linked, making them a type of organic halides. They are known under many chemical names and are widely used as fire extinguishers, propellants, and solvents. Some haloalkanes have negative effects on the environment such as ozone depletion. Uses of certain chloroalkanes have been phased out by the IPPC directive on greenhouse gases in 1994 and by the Volatile Organic Compounds (VOC) directive of the EU (European Union) in 1997.

2. Ozone depletion:

The ozone layer (ozonosphere layer) is part of the earth's atmosphere, which contains relatively high concentrations of ozone. The ozone layer was first discovered by two French physicists, Charles Fabry and Henry Buisson, in 1913.

About 90% of the ozone in our atmosphere is contained in the stratosphere, the region from about 10 to 50 km above the earth's surface. The remaining 10% of the ozone is contained in the troposphere, the lowest part of our atmosphere. Although the concentration of ozone in the ozone layer is very small, it is vitally important to life because it absorbs biologically harmful ultraviolet radiation (UV) emitted from the sun. Depletion of the ozone layer by VOCs allows more of the harmful UV to reach the earth and causing increased genetic damage to living things.

3. Insecticides:

An insecticide is a pesticide used against insects in all developmental forms. Insecticides are widely used in agriculture, households, businesses, and even in medicine. Insecticides are one of the factors behind the increase in the agricultural productivity of the 20th century. It is very important to balance agricultural needs with safety concerns because the insecticides have the potential to significantly alter the ecosystems by entering the food chain or devastating populations of helpful insects such as honeybees. Honeybees are responsible for the pollination that results in multiple important food crops.

4. Pollution events:

Pollution of the air, water, and soil occurs in many ways, including smokestack output, heat pollution of rivers from manufacturing plants, plastic pollution of the ocean, toxic metal runoff into streams from coal mining, radioactivity releases from nuclear power plant accidents, acid rain pollution of soils, and the addition of excessive carbon dioxide to the atmosphere which causes increased greenhouse warming and climate changes. There have been many pollution events in our history. One was the oil pollution in 1971 during the Gulf war in Kuwait. This single event affected marine life and birds. The oil spill on a large scale gave no chance for the birds to survive. It was graphic, expensive, and an ecological disaster. Other oil spill disasters include the tanker Valdez incident and the Deep Horizon BP spill in the Gulf of Mexico in 2010. Receiving less attention than dramatic spills are the daily oil leaks in small and steady amounts from pipelines and tankers and trucks as they transport oil.

Skill 2.4   Analyze the synergistic relationship between basic and applied research, technology, the economy and public welfare.

A synergistic relationship can be defined as a relationship in which the outcomes of the components working together is greater than the sum of the outcomes of the components working separately. In science, basic and applied research, technology, economy, and the public welfare all interact to drive advancement in science.. This relationship is about intent and commonness. It is also a kind of interdependence among all those that are involved or participating.

Some very important qualities / characteristics of synergistic relationship are:

1. Open communication
2. Win-Win situation
3. Explore all options

4. Well defined objectives and realistic expectations
5. Clear roles and responsibilities
6. Take and share risks
7. Expect the unexpected and be flexible

There could be many more, but these are some of the important qualities.

The relationship among the basic and applied research, technology, economy and public or the society has to be synergistic. Basic research is the starting point in this chain of events. Basic research provides with knowledge, which is again of two types. The first type of knowledge is theoretical knowledge, giving us the understanding of processes. Whereas the second type of knowledge could be applied for the benefit of humanity / society / public. Applied research is great value because it is directly useful to us; it deals with issues like Aids, Tuberculosis, HPV, Parkinson's etc., to name a few. This is very important to society because it is useful to the society / public. Public is interested in it and public has its opinion about the research. Let's look at stem cell research for an example. There are people for and against this controversial piece of research. We are living in the age of technology. Our lives are so intertwined, dependent and in tune with technology. We are afraid that we may not be able to function as human beings any longer without technology. Such is the relationship of public with technology. The relationship of public with technology sometimes looks dangerous. Whatever technology offers us on a platter, we are so ready to use it whatever it might be. The economy is like the blood in the veins, without which we are dead and gone. The economy, technology and public are inseparable, in that our money, comforts and modern knowledge are so intertwined with each of these three fields.

This synergistic relationship overlaps some moral and ethical issues. Whatever research is done, public has a right to know it. In a synergistic relationship, as mentioned earlier, there has to be open communication of what is going on.

The economy should not be the guiding principle for any piece of research or technology. Clear objectives are absolutely important because the public has a stake in all these ventures, especially if they are federally funded. If they are privately funded, the organizations need to remember that they are bound by social ethics and correct practices.

The ultimate goal of any research, technology project or economic venture must be for the benefit of the public and there should be clear cut objectives and the flexibility to expect the unexpected and the ability to deal with it.

Skill 2.5 **Analyze the causes and effects of multidrug resistance and globalization on the spread and treatment of human pathogens.**

Multidrug resistance (MDR) is the ability of pathologic cells to show resistance to a wide variety of structurally and functionally unrelated compounds. These pathologic cells include bacterial, fungal and neoplastic (tumor) cells.

The widespread occurrence of MDR (multidrug resistance) in tumor cells presents a major obstacle to successful cancer chemotherapy. Neoplastic resistance is the term used to describe MDR in tumor cells. In cultured cancer cells, there are increased levels of a glycoprotein called P-glycoprotein (P-gp). It is an ATP dependent extrusion pump for drugs or any other substrates.

The DNA of cancer cells has been modified such that the cells grow at an abnormal rate, unable to perform their function and in many cases they damage the surrounding cells. The genes involved in this abnormal growth of cancer cells are ras (provides energy for cell division) and p53 (like a brake which stops cell division). When these two are damaged, the cells keep on dividing without control.

The mechanism of MDR in cancer cells is as follows:

- Increased efflux of drugs (by P-gp, P-glycoprotein)
- Enzymatic deactivation of antibiotics (i.e. glutathione conjugation)
- Decreased permeability of the cells (drugs can't get in)
- Altered binding sites
- Alternate metabolic pathways

There are other human pathogens like TB, HIV, Salmonella, Candida, etc. which have become multidrug resistant, causing huge problems to our health. The mechanism of MDR is almost the same in all of these human pathogens. It is important to know the spread of these human pathogens. In this day and age, with increased communication and travel, there is a huge change in the lifestyles of people.

- Outdoor lifestyle: As we play and move around outside, we are at greater risk of being bitten by mosquitoes, ticks and other insects, which carry diseases like West Nile virus, Ehlrichia etc.
- Tourism: Tourism is a very popular now. The inner curiosity of man to explore and visit exotic, historical, and recreational spots is coming with a price. When people visit countries that have some kinds of diseases, there is a very strong possibility that some will be infected, though they had precautions (e.g., malaria from Caribbean islands and other tropics).
- Moist, humid environment: Wherever there is this type of environment, we can expect molds, especially black mold, which is not healthy.
- Cruise ships: The luxurious cruise ships have a downside to them. They have a virus called Norwalk Virus, which survives in enclosed atmospheres like that of a cruise ship.

- Some rural communities: Florida's rural communities are facing the problem of an increase in TB and HIV cases and also Cryptococcus.
- Migratory birds: Birds migrate for various reasons, but they also carry avian flu. Avian Influenza is proving deadly of late.

The picture of multidrug resistance looks very dim. But researchers are trying to find solutions to this gigantic problem. Research is being done on the development of drugs that either avoid or inhibit these efflux pumps (ABC & MFS), allowing medication to slip inside the bacterial cells and kill them. Some researchers are concentrating on disabling the multidrug efflux pump.

Some important aspects of the treatment of MDR:

1. Identification of the pathogen
2. Using a specific antibiotic for that pathogen rather than broad spectrum agents
3. Using the full course of antibiotic as prescribed by the physician
4. Not to use antibiotics for viral infections like cold, cough, bronchitis, etc.

Skill 2.6   Demonstrate knowledge of pertinent legislation and national guidelines (e.g., NABT, ISEF) regarding laboratory safety, hazardous materials, experimentation, and/or the use of organisms in the classroom

In addition to requirements set forth by your place of employment, the NABT (National Association of Biology Teachers) and ISEF (International Science Education Foundation) have been instrumental in setting parameters for the science classroom. All science laboratories should contain the following items of safety equipment.

The following are requirements by law.

- Fire blanket which is visible and accessible
- Ground Fault Circuit Interrupters (GFCI) within two feet of water supplies signs designating room exits
- Emergency shower providing a continuous flow of water
- Emergency eye wash station which can be activated by the foot or forearm
- Eye protection for every student and a means of sanitizing equipment
- Emergency exhaust fans providing ventilation to the outside of the building
- Master cut-off switches for gas, electric and compressed air. Switches must have permanently attached handles. Cut-off switches must be clearly labeled.
- An ABC fire extinguisher
- Storage cabinets for flammable materials

Also recommended, but not required by law:

- Chemical spill control kit

- Fume hood with a motor which is spark proof
- Protective laboratory aprons made of flame retardant material
- Signs which will alert potentially hazardous conditions
- Containers for broken glassware, flammables, corrosives and waste.
- Containers should be labeled.

Students should wear safety goggles when performing dissections, heating, or while using acids and bases. Hair should always be tied back and objects should never be placed in the mouth. Food should not be consumed while in the laboratory. Hands should always be washed before and after laboratory experiments. In case of an accident, eye washes and showers should be used for eye contamination or a chemical spill that covers the student's body. Small chemical spills should only be contained and cleaned by the teacher. Kitty litter or a chemical spill kit should be used to clean spills. For large spills, the school administration and the local fire department should be notified. Biological spills should also only be handled by the teacher. Contamination with biological waste can be cleaned by using bleach when appropriate. Accidents and injuries should always be reported to the school administration and local health facilities. The severity of the accident or injury will determine the course of action to pursue.

It is the responsibility of teachers to provide a safe environment for the students. Proper supervision greatly reduces the risk of injury and a teacher should never leave a class for any reason without providing alternate supervision. After an accident, two factors are considered: foreseeability and negligence.

Foreseeability is the anticipation that an event may occur under certain circumstances. **Negligence** is the failure to exercise ordinary or reasonable care. Safety procedures should be a part of the science curriculum and a well-managed classroom is important to avoid potential lawsuits. The **"Right to Know Law" statutes** cover science teachers who work with potentially hazardous chemicals. Briefly, the law states that employees must be informed of potentially toxic chemicals. An inventory must be made available if requested. The inventory must contain information about the hazards and properties of the chemicals. Training must be provided in the safe handling and interpretation of the Material Safety Data Sheet.

The following chemicals are potential carcinogens and are not allowed in school facilities: Acrylonitrile, Arsenic compounds, Asbestos, Benzedrine, Benzene, Cadmium compounds, Chloroform, Chromium compounds, Ethylene oxide, Ortho-toluidine, Nickel powder, Mercury.

Chemicals should not be stored on bench tops or heat sources. They should be stored in groups based on their reactivity with one another and in protective storage cabinets. All containers within the lab must be labeled. Suspected and known carcinogens must be labeled as such and segregated within trays to contain leaks and spills. Chemical waste should be disposed of in properly labeled containers. Waste should be separated based on their reactivity with other chemicals.

Material safety data sheets are available for every chemical and biological substance. These are available directly from the company of acquisition or the internet. The manuals for equipment used in the lab should be read and understood before using them.

All laboratory solutions should be prepared as directed in the lab manual. Care should be taken to avoid contamination. All glassware should be rinsed thoroughly with distilled water before using and cleaned well after use. All solutions should be made with distilled water as tap water contains dissolved particles that may affect the results of an experiment. Unused solutions should be disposed of according to local disposal procedures.

Biological material should never be stored near food or water used for human consumption. All biological material should be appropriately labeled. All blood and body fluids should be put in a well-contained container with a secure lid to prevent leaking. All biological waste should be disposed of in biological hazardous waste bags.

Use of live specimens

No dissections may be performed on living mammalian vertebrates or birds. Lower order life and invertebrates may be used. Biological experiments may be done with all animals except mammalian vertebrates or birds. No physiological harm may result to the animal. All animals housed and cared for in the school must be handled in a safe and humane manner. Animals are not to remain on school premises during extended vacations unless adequate care is provided. Any instructor who intentionally refuses to comply with the laws may be suspended or dismissed.

Pathogenic organisms must never be used for experimentation. Students should adhere to the following rules at all times when working with microorganisms to avoid accidental contamination:

1. Treat all microorganisms as if they were pathogenic.
2. Maintain sterile conditions at all times

Animals which are not obtained from recognized sources should not be used. Decaying animals or those of unknown origin may harbor pathogens and/or parasites. Specimens should be rinsed before handling. Latex gloves are desirable. If not available, students with sores or scratches should be excused from the activity. Formaldehyde is likely carcinogenic and should be avoided or disposed of according to district regulations. Students objecting to dissections for moral reasons should be given an alternative assignment. Interactive dissections are available online or from software companies for those students who object to performing dissections. There should be no penalty for those students who refuse to physically perform a dissection.

Competency 3.0   Knowledge of the Chemical Processes of Living Things

Skill 3.1   Identify the structures, functions, and importance of inorganic and organic compounds (e.g., water, mineral salts, carbohydrates, lipids, proteins, nucleic acids) in cells.

Water is necessary for life. Its properties are due to its molecular structure and it is an important solvent in biological compounds. Water is a polar substance. This means it is formed by covalent bonds that make it electrically lopsided. Water molecules are attracted to other water molecules due to this electrical attraction and allow for two important properties: **adhesion** and **cohesion**.

Adhesion is when water sticks to other substances like the xylem of a stem, which aids the water in traveling up the stem to the leaves.

Cohesion is the ability of water molecules to stick to each other by hydrogen bonds. This allows for surface tension on a body of water or capillarity, which allows water to move through vessels. Surface tension is how difficult it is to stretch or break the surface of a liquid. Cohesion allows water to move against gravity.

There are several other important properties of water. Water is a good solvent. An aqueous solution is one in which water is the solvent. It provides a medium for chemical reactions to occur. Water has a high specific heat of 1 calorie per gram per degree Celsius, allowing it to cool and warm slowly, allowing organisms to adapt to temperature changes.

Water has a high boiling point it is a good coolant. Its ability to evaporate stabilizes the environment and allows organisms to maintain body temperature. Water is most dense at four degrees centigrade. Water has a high freezing point and a lower density as a solid than as a liquid. This allows ice to float on top of water so a whole body of water does not freeze during the winter. In this way, animals may survive the winter.

A compound consists of two or more elements. There are four major chemical compounds found in the cells and bodies of living things. These include carbohydrates, lipids, proteins and nucleic acids.

Monomers are the simplest unit of structure. **Monomers** can be combined to form **polymers**, or long chains, making a large variety of molecules possible. Monomers combine through the process of condensation reaction (also called dehydration synthesis). In this process, a hydrogen ion from one reactant and a hydroxyl ion from the other reactant are removed and joined together to form one molecule of water being removed between each of the adjoining molecules. In order to break the molecules apart in a polymer, water molecules are added between monomers in the reverse of the above process, thus breaking the bonds between them. This is called hydrolysis.

Carbohydrates contain a ratio of two hydrogen atoms for each carbon and oxygen $(CH_2O)_n$. Carbohydrates include sugars and starches. They function in the release of energy. **Monosaccharides** are the simplest sugars and include glucose, fructose, and galactose. They are major nutrients for cells. In cellular respiration, the cells extract the energy in glucose molecules. **Disaccharides** are made by joining two monosaccharides by condensation to form a glycosidic linkage (covalent bond between two monosaccharides). Maltose is formed from the combination of two glucose molecules, lactose is formed from joining glucose and galactose, and sucrose is formed from the combination of glucose and fructose.

Glucose   Galactose   Fructose

Sucrose

Polysaccharides consist of many monomers joined. They are storage material hydrolyzed as needed to provide sugar for cells or building material for structures protecting the cell. Examples of polysaccharides include starch, glycogen, cellulose and chitin.

   Starch - major energy storage molecule in plants. It is a polymer consisting of glucose monomers.

   Glycogen - major energy storage molecule in animals. It is made up of many glucose molecules.

   Cellulose - found in plant cell walls, its function is structural. Many animals lack the enzymes necessary to hydrolyze cellulose, so it simply adds bulk (fiber) to the diet.

   Chitin - found in the exoskeleton of arthropods and fungi. Chitin contains an amino sugar (glycoprotein).

Phospholipids are a vital component in cell membranes. In a phospholipid, a phosphate group linked to a nitrogen group replaces one or two fatty acids. They consist of a **polar** (charged) head that is hydrophilic or water loving and a **nonpolar** (uncharged) tail which is hydrophobic or water fearing. This allows the membrane to orient itself with the polar heads facing the interstitial fluid found outside the cell and the internal fluid of the cell.

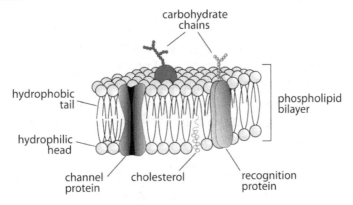

Lipids are composed of glycerol (an alcohol) and three fatty acids. Lipids are **hydrophobic** (water fearing) and will not mix with water. There are three important families of lipids, fats, phospholipids and steroids.

Fats consist of glycerol (alcohol) and three fatty acids. Fatty acids are long carbon skeletons. The nonpolar carbon-hydrogen bonds in the tails of fatty acids are why they are hydrophobic. Fats are solids at room temperature and come from animal sources (butter, lard).

Steroids are insoluble and are composed of a carbon skeleton consisting of four inter-connected rings. An important steroid is cholesterol, which is the precursor from which other steroids are synthesized. Hormones, including cortisone, testosterone, estrogen, and progesterone, are steroids. Their insolubility keeps them from dissolving in body fluids.

Proteins compose about fifty percent of the dry weight of animals and bacteria. Proteins function in structure and aid in support (connective tissue, hair, feathers, quills), storage of amino acids (albumin in eggs, casein in milk), transport of substances (hemoglobin), hormonal to coordinate body activities (insulin), membrane receptor proteins, contraction (muscles, cilia, flagella), body defense (antibodies), and as enzymes to speed up chemical reactions.

All proteins are made of long chains of **amino acids**. There are twenty different amino acids, and a typical protein may be hundreds of amino acids long. An amino acid contains an amino group and an acid group. The radical group varies and defines the amino acid. Amino acids form through condensation reactions with the removal of water. The bond that is formed between two amino acids is called a peptide bond. Polymers of amino acids are called polypeptide chains. An analogy can be drawn between the twenty amino acids and the alphabet.

Millions of words can be formed using an alphabet of only twenty-six letters. This diversity is also possible using only twenty amino acids. This results in the formation of many different proteins, whose structure defines the function.

There are four levels of protein structure: primary, secondary, tertiary, and quaternary.

Primary structure is the protein's unique sequence of amino acids. A slight change in primary structure can affect a protein's conformation and its ability to function. **Secondary structure** is the coils and folds of polypeptide chains. The coils and folds are the result of hydrogen bonds along the polypeptide backbone. The secondary structure is either in the form of an alpha helix or a pleated sheet. The alpha helix is a coil held together by hydrogen bonds. A pleated sheet is the polypeptide chain folding back and forth. The hydrogen bonds between parallel regions hold it together. **Bonding between the side chains of the amino acids forms tertiary structure**. Disulfide bridges are created when two sulfhydryl groups on the amino acids bond together to form a strong covalent bond. **Quaternary structure** is the overall structure of the protein from the aggregation of two or more polypeptide chains. An example of this is hemoglobin. Hemoglobin consists of two kinds of polypeptide chains.

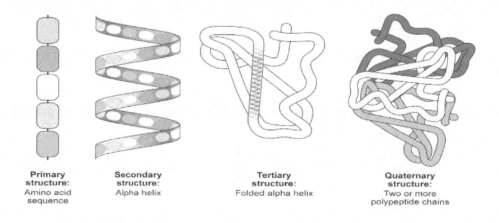

Primary
structure:
Amino acid
sequence

Secondary
structure:
Alpha helix

Tertiary
structure:
Folded alpha helix

Quaternary
structure:
Two or more
polypeptide chains

Nucleic acids consist of DNA (deoxyribonucleic acid) and RNA (ribonucleic acid). Nucleic acids contain the instructions for the amino acid sequence of proteins and the instructions for replicating. The monomer of nucleic acids is called a nucleotide. A nucleotide consists of a 5-carbon sugar, (deoxyribose in DNA, ribose in RNA), a phosphate group, and a nitrogenous base. The base sequence codes for the instructions. There are five bases: adenine, thymine, cytosine, guanine, and uracil. Uracil is found only in RNA and replaces the thymine. A summary of nucleic acid structure can be seen in the table below:

	SUGAR	PHOSPHATE	BASES
DNA	Deoxyribose	Present	adenine, **thymine**, cytosine, guanine
RNA	Ribose	Present	adenine, **uracil**, cytosine, guanine

Due to the molecular structure, adenine will always pair with thymine in DNA or uracil in RNA. Cytosine always pairs with guanine in both DNA and RNA. This allows for the symmetry of the DNA molecule seen below.

DNA

sugar-phosphate = `backbone'

nitrogenous bases

phosphate molecule   weak hydrogen bonds   deoxyribose

Adenine and thymine (or uracil) are linked by two covalent bonds and cytosine and guanine are linked by three covalent bonds. The guanine and cytosine bonds are harder to break apart than thymine (uracil) and adenine because of the greater number of these bonds. The DNA molecule is called a double helix due to its twisted ladder shape.

Skill 3.2   Compare and apply the laws of thermodynamics to living systems, including the role of enzymes in biological reactions.

The first and second laws of thermodynamics have implications for living systems. The first states that energy is not created or destroyed though it can change form. In living systems energy changes from sunlight to chemical bond energy of glucose, then to the chemical bond energy of ATP, then is released to fuel a variety of life processes. These energy transformations and life processes are mediated by enzymes, which are catalysts; that is, they allow the processes to occur with smaller inputs of energy.

Enzymes act as biological catalysts to speed up reactions. Enzymes are the most diverse of all types of proteins. They are not used up in a reaction and are recyclable. Each enzyme is specific for a single reaction. Enzymes act on a substrate.

The substrate is the material to be broken down or put back together. Most enzymes end in the suffix -ase (lipase, amylase). The prefix is the substrate being acted on (lipids, sugars).

The active site is the region of the enzyme that binds to the substrate. There are two theories for how the active site functions. The **lock and key theory** states that the shape of the enzyme is specific because it fits into the substrate like a key fits into a lock. It aids in holding molecules close together so reactions can easily occur. The **Induced fit theory** states that an enzyme can stretch and bend to fit the substrate. This is the most accepted theory.

The second law of thermodynamics states that systems move toward increasing levels of entropy, which means more energy in the form of thermal energy (heat). Thermal energy is not an efficient energy type for doing work, including biological work. Therefore, living systems need a constant input of energy, which is light from the sun.

Skill 3.3   Predict the effects of changes in pH, temperature, substrate concentration, and enzyme concentration on enzyme activity.

Many factors can affect enzyme activity. Temperature and pH are two of those factors. The temperature can affect the rate of reaction of an enzyme. The optimal pH for most enzymes is between 6 and 8, with a few enzymes whose optimal pH falls out of this range, such as those that work in the stomach. Temperature also impacts enzyme activity, with high temperatures changing the structure of enzymes and completely deactivating them, and low temperatures slowing the molecular movement that facilitates reactions. An increase in substrate concentration would increase enzyme activity up to the point where enzymes molecules are active, and an increase in enzyme concentration would increase enzyme activity up to the point that all substrate molecules are engaged by enzymes.

Cofactors aid in the enzyme's function. Cofactors may be inorganic or organic. Organic cofactors are known as coenzymes. An example of a coenzyme is vitamins. Some chemicals can inhibit an enzyme's function. **Competitive inhibitors** block the substrate from entering the active site of the enzyme to reduce productivity. **Noncompetitive inhibitors** bind to the enzyme in a location not in the active site but still interrupt substrate binding. In most cases, noncompetitive inhibitors alter the shape of the enzyme. An **allosteric enzyme** can exist in two shapes; they are active in one form and inactive in the other. Overactive enzymes may cause metabolic diseases.

 **Skill 3.4** **Identify substrates, products, and relationships between glycolysis, Krebs cycle, and electron transport, including the respiration of carbohydrates, fats, and amino acids.**

Cellular respiration is the metabolic pathway in which food (glucose, etc.) is broken down to release energy from the chemical bonds of glucose and transfer it to the chemical bonds of ATP. Both plants and animals utilize respiration to release energy for metabolism. In respiration, energy is released by the transfer of electrons in a process known as an **oxidation-reduction (redox)** reaction. The oxidation phase of this reaction is the loss of an electron and the reduction phase is the gain of an electron. Redox reactions are important for the stages of respiration.

Glycolysis is the first step in respiration. It occurs in the cytoplasm of the cell and does not require oxygen. A specific enzyme catalyzes each of the ten stages of glycolysis.

The following is a summary of those stages. In the first stage, the reactant is glucose. For energy to be released from glucose, it must be converted to a reactive compound. This conversion occurs through the phosphorylation of a molecule of glucose by the use of two molecules of ATP. This is an investment of energy by the cell. The six-carbon product, called fructose -1,6- bisphosphate, breaks into two 3-carbon molecules of sugar. A phosphate group is added to each sugar molecule and hydrogen atoms are removed. Hydrogen is picked up by NAD^+ (a vitamin). Since there are two sugar molecules, two molecules of NADH are formed. The reduction (adding of hydrogen) of NAD allows the potential of energy transfer. As the phosphate bonds are broken, ATP is made. Two ATP molecules are generated as each original 3-carbon sugar molecule is converted to pyruvic acid (pyruvate). A total of four ATP molecules are made in the four stages. Since two molecules of ATP were needed to start the reaction in stage 1, there is a net gain of two ATP molecules at the end of glycolysis. This accounts for only two percent of the total energy in a molecule of glucose.

Beginning with pyruvate, which was the end product of glycolysis, the following steps occur before entering the **Krebs cycle**.

1. Pyruvic acid is changed to acetyl-CoA (coenzyme A). This is a three-carbon pyruvic acid molecule, which has lost one molecule of carbon dioxide (CO_2) to

become a two-carbon acetyl group. Pyruvic acid loses a hydrogen to NAD^+ which is reduced to NADH.

2. Acetyl CoA enters the Krebs cycle. For each molecule of glucose it started with, two molecules of Acetyl CoA enter the Krebs cycle (one for each molecule of pyruvic acid formed in glycolysis)

The **Krebs cycle** (also known as the citric acid cycle), occurs in four major steps. First, the two-carbon acetyl CoA combines with a four-carbon molecule to form a six-carbon molecule of citric acid. Next, two carbons are lost as carbon dioxide (CO_2) and a four-carbon molecule is formed to become available to join with CoA to form citric acid again. Since we started with two molecules of CoA, two turns of the Krebs cycle are necessary to process the original molecule of glucose. In the third step, eight hydrogen atoms are released and picked up by FAD and NAD (vitamins and electron carriers).

Lastly, for each molecule of CoA (remember there were two to start with) you get:

3 molecules of NADH x 2 cycles
1 molecule of $FADH_2$ x 2 cycles
1 molecule of ATP x 2 cycles

Therefore, this completes the breakdown of glucose.

At this point, a total of four molecules of ATP have been made; two from glycolysis and one from each of the two turns of the Krebs cycle. Six molecules of carbon dioxide have been released; two prior to entering the Krebs cycle, and two for each of the two turns of the Krebs cycle. Twelve carrier molecules have been made; ten NADH and two $FADH_2$. These carrier molecules will carry electrons to the electron transport chain. ATP is made by substrate level phosphorylation in the Krebs cycle. Notice that the Krebs cycle in itself does not produce much ATP, but functions mostly in the transfer of electrons to be used in the electron transport chain where the most ATP is made.

In the **Electron Transport Chain,** NADH transfers electrons from glycolysis and the Krebs cycle to the first molecule in the chain of molecules embedded in the inner membrane of the mitochondrion. Most of the molecules in the electron transport chain are proteins. Nonprotein molecules are also part of the chain and are essential for the catalytic functions of certain enzymes. The electron transport chain does not make ATP directly. Instead, it breaks up a large free energy drop into a more manageable amount. The chain uses electrons to pump H^+ across the mitochondrion membrane. The H^+ gradient is used to form ATP synthesis in a process called **chemiosmosis** (oxidative phosphorylation). ATP synthetase and energy generated by the movement of hydrogen ions coming off of NADH and $FADH_2$ builds ATP from ADP on the inner membrane of the mitochondria. Each NADH yields three molecules of ATP (10 x 3) and each $FADH_2$ yields two molecules of ATP (2 x 2). Thus, the electron transport chain and oxidative phosphorylation produces 34 ATP.

So, the net gain from the whole process of respiration is 36 molecules of ATP:

Process	# ATP produced (+)	# ATP consumed (-)	Net # ATP
Glycolysis	4	2	+2
Acetyl CoA	0	2	-2
Krebs cycle	1 per cycle (2 cycles)	0	+2
Electron transport chain	34	0	+34
Total			+36

Skill 3.5 **Compare end products and energy yields of alcoholic fermentation, lactic acid fermentation, and aerobic respiration.**

Glycolysis generates ATP with oxygen (aerobic) or without oxygen (anaerobic). Aerobic respiration has already been discussed. Anaerobic respiration can occur by fermentation. ATP can be generated by fermentation by substrate level phosphorylation if there is enough NAD^+ present to accept electrons during oxidation. In anaerobic respiration, NAD^+ is regenerated by transferring electrons to pyruvate. There are two common types of fermentation.

In **alcoholic fermentation**, pyruvate is converted to ethanol in two steps. In the first step, carbon dioxide is released from the pyruvate. In the second step, ethanol is produced by the reduction of acetaldehyde by NADH. This results in the regeneration of NAD^+ for glycolysis. Yeast and some bacteria carry out alcohol fermentation.

NADH reduces pyruvate to form lactate as a waste product in the process of **lactic acid fermentation**. Animal cells and some bacteria that do not use oxygen utilize lactic acid fermentation to make ATP. Lactic acid forms when pyruvic acid accepts hydrogen from NADH. A buildup of lactic acid is what causes muscle soreness following exercise.

Energy remains stored in the lactic acid or alcohol until needed. This is not an efficient type of respiration. When oxygen is present, aerobic respiration occurs after glycolysis.

Both aerobic and anaerobic pathways oxidize glucose to pyruvate by glycolysis and both pathways have NAD^+ as the oxidizing agent. A substantial difference between the two pathways is that in fermentation, an organic molecule such as pyruvate or acetaldehyde is the final electron acceptor. In respiration, the final electron acceptor is oxygen. Another key difference is that respiration yields much more energy from a sugar molecule than fermentation does. Respiration can produce up to 18 times more ATP than fermentation.

Skill 3.6 Identify the raw materials and products of C-3 photosynthesis, including the Calvin cycle, light dependent and light independent reactions, and factors that affect their rate.

The formula for photosynthesis is:

$$CO_2 + H_2O + \text{energy (from sunlight)} \rightarrow C_6H_{12}O_6 + O_2$$

Photosynthesis reverses the electron flow. Water is split by the chloroplast into hydrogen and oxygen. The oxygen is given off as a waste product as carbon dioxide is reduced to sugar (glucose). This requires the input of energy, which comes from the sun. The sunlight energy is being transferred to the chemical bond energy holding the carbon atoms together in the glucose molecule.

Photosynthesis occurs in two stages: the light reactions and the Calvin cycle (dark reactions). The conversion of solar energy to chemical energy occurs in the light reactions. Electrons are transferred by the absorption of light by chlorophyll and cause the water to split, releasing oxygen as a waste product. The chemical energy that is created in the light reaction is carried in the chemical bonds of NADPH. ATP is also produced by a process called photophosphorylation; ATP has high energy bonds. These energy-carrying molecules are produced in the thylakoids and are used in the Calvin cycle to produce sugar.

The second stage of photosynthesis is the **Calvin cycle**. Carbon dioxide in the air is incorporated into organic molecules already in the chloroplast. The NADPH produced in the light reaction is used as reducing power for the reduction of the carbon to carbohydrate. ATP from the light reaction is also needed to convert carbon dioxide to carbohydrate (sugar).

The two stages of photosynthesis are summarized below.

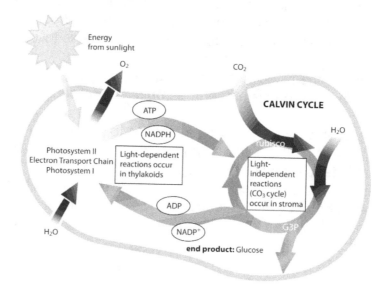

The process of photosynthesis is made possible by the presence of the sun. Visible light ranges in wavelengths of 750 nanometers (red light) to 380 nanometers (violet light). As wavelength decreases, the amount of available energy increases. Light is carried as photons, which is a fixed quantity of energy. Light is reflected (what we see), transmitted, or absorbed (what the plant uses). The plant's pigments capture light of specific wavelengths. Remember that the light that is reflected is what we see as color.

Plant pigments include:

Chlorophyll *a* - reflects green/blue light; absorbs red light
Chlorophyll *b* - reflects yellow/green light; absorbs red light
Carotenoids - reflects yellow/orange; absorbs violet/blue

The pigments absorb photons. The energy from the light excites electrons in the chlorophyll that jump to orbitals with more potential energy and reach an "excited" or unstable state.

The high-energy electrons are trapped by primary electron acceptors, which are located on the thylakoid membrane. These electron acceptors and the pigments form reaction centers called photosystems that are capable of capturing light energy. Photosystems contain a reaction-center chlorophyll that releases an electron to the primary electron acceptor. This transfer is the first step of the light reactions. There are two photosystems, named according to their date of discovery, not their order of occurrence.

Photosystem I is composed of a pair of chlorophyll *a* molecules. Photosystem I is also called P700 because it absorbs light of 700 nanometers. Photosystem I makes ATP whose energy is needed to build glucose.

Photosystem II - this is also called P680 because it absorbs light of 680 nanometers. Photosystem II produces ATP + NADPH$_2$ and the waste gas oxygen.
Both photosystems are bound to the **thylakoid membrane**, close to the electron acceptors.

The production of ATP is termed **photophosphorylation** due to the use of light. Photosystem I uses cyclic photophosphorylation because the pathway occurs in a cycle. It can also use noncyclic photophosphorylation, which starts with light and ends with glucose. Photosystem II uses noncyclic photophosphorylation only.

Both photosystems are bound to the **thylakoid membrane**, close to the electron acceptors.

Skill 3.7   Identify key differences between C-3, C-4, and CAM photosynthesis, and the ecological significance of these pathways

Depending on the first product formed during photosynthesis, scientists have identified three types of plants – C3, C4 and CAM (Crassulacean Acid Metabolism).

C3 plants: The first product of photosynthesis in these plants is a 3 - carbon compound known as 3-phosphoglycerate. In C3 - plants carbon dioxide is assimilated and is directly fed into the Calvin cycle. The chloroplasts of C3 cells are of homogeneous structure.

C4 plants: The assimilated carbon dioxide is bound to phosphoenolpyruvate (PEP) in mesophyll cells. The first product of photosynthesis in this C4 cycle or Hatch – Slack cycle is oxaloacetate, which is produced by the addition of one molecule of carbon dioxide to phosphoenolpyruvate (PEP). The common examples of this C4 cycle plants are Graminae like maize or sugar cane. The anatomies of C3 and C4 cycles differ. There are two types of chloroplasts in C4 plants. The mesophyll cells contain normal chloroplasts and the vascular bundle sheaths have chloroplasts without grana i.e. they are partially impaired in function. This anomaly affects only the light reactions, not the dark reactions.

CAM or Crassulacean Acid Metabolism: CAM is the abbreviation of Crassulacean Acid Metabolism. As the name suggests, this pathway occurs only in Crassulacean species and other succulent species. The chemical reaction of the carbon dioxide accumulation is similar to that of C4 plants, but here the carbon dioxide fixation and assimilation are not separated spatially, but in time. There is a time lapse between the two processes. CAM plants are found mostly in arid/desert regions. The opening of stomata results usually in a large loss of water. To avoid this loss of water, these plants have a mechanism, which allows them to fix carbon dioxide during the night, and the prefixed carbon dioxide is stored in the vacuoles as malate and isocitrate and is utilized in the daytime during photosynthesis.

The enzyme that catalyzes the primary carbon dioxide fixation of C4 and CAM plants is phosphoenolpyruvate carboxylase (PEPC). It is much more efficient in fixing carbon dioxide than Rubisco, the first enzyme of the Calvin cycle. As a result, C4 plants are able to utilize even trace amounts of carbon dioxide. This enzyme, PEPC also occurs in small quantities (about 3%) in C3 plants and also plays a key position in the metabolic regulation.

In the temperate regions of the world, the low light intensity is a disadvantage for the C4 plants. The net rate of photosynthesis is more in higher light intensities than the C3 plants. C3 plants have an advantage over C4 and CAM plants due to their low rate of photorespiration and because they do not need energy for fixing carbon dioxide earlier.

In low temperatures and where light is a limiting factor, C3 have advantage over C4 plants. C4 plants are found only in warmer temperatures and they are either shrubs or herbs. Spartina townsendii, a C4 plant, is an exception to this rule. C4 plants have a mixture of unrelated monocotyledons and dicotyledons as well as Anacystis nidulans, a blue green alga and some dinoflagellates.

The most important thing here is there is a genetic potential for both C3 and C4 pathways is quite common in the plant kingdom and depending on the ecological needs of the plants, one pathway is chosen by a species and another pathway by a related species. In the genus, Atriplex, a study showed that both pathways occur. In several species of Zea, Mollugo, Moricandia, Flaveria, and others, both types of carbon dioxide fixation occur in the same plant. In younger plants, the C3 pathway is used and in the older ones, the C4 pathway is taken.

CAM advantages and disadvantages: CAM has been detected in more than 1000 angiosperms belonging to 17 different families. It usually occurs in succulent plants, but succulence is not a precondition for CAM. Some non-succulent plants use Cam and some succulent plants switch to CAM when growing in saline soils. Depending on the environment, some plants switch to CAM for survival.

Skill 3.8 Identify and analyze the process of chemiosmosis in photosynthesis and respiration

Chemiosmosis is also known as a facilitated diffusion. Chemiosmosis requires a phospholipids bilayer, a proton pump, protons and ATPase.

Chemical energy is used to pump protons through a proton pump (intrinsic protein). This creates a high concentration of protons ($H+$) inside the thylakoid disk. ATPase has a channel that allows for the facilitated diffusion of protons back in through the membrane. This activates ATPase, which in turn catalyzes the formation of ATP.

This process can be summarized as follows:

The chemical energy is transferred to a proton pump, which creates a proton gradient across the membrane. This gradient is used to activate the enzyme ATPase. ATPase in turn catalyzes the formation of ATP.

Chemiosmotic mechanism in photosynthesis:

1. High energy electrons are created upon absorption of a photon
2. These are then passed through an electron transport system which then affects the pumping of photons
3. ATP is synthesized from ADP and inorganic phosphate through chemiosmosis.

Chemiosmosis in respiration:

1. The energy released during respiration as electrons pass down the gradient from NADH to oxygen is being utilized by three enzyme complexes of the respiratory chain to pump protons (H+) against their concentration gradient from the matrix of the mitochondrion into the inter membrane space.
2. As their concentration increases there, a diffusion gradient is set up.
3. The only way out for these protons is through the ATP synthase complex.
4. As in chloroplasts during photosynthesis, the energy released as these protons flow down their gradient is used to synthesize ATP. This is an example of facilitated diffusion.

Below is a diagram of the relationship between photosynthesis and cellular respiration.

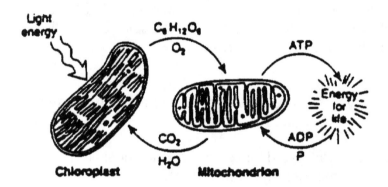

Skill 3.9 Compare heterotrophy and autotrophy and the roles of these processes in the environment

An **autotroph** (self feeder) is an organism that makes its own food from the energy of the sun or other elements. Autotrophs include:

1. **photoautotrophs** - make food from light and carbon dioxide releasing oxygen that can be used for respiration.
2. **chemoautotrophs** - oxidize sulfur and ammonia; this is done by some bacteria.

Heterotrophs (other feeder) are organisms that must eat other living things for their energy. **Consumers** are the same as heterotrophs; all animals are heterotrophs. **Decomposers** break down once living things. Bacteria and fungi are examples of decomposers. **Scavengers** eat dead things. Examples of scavengers are bacteria, fungi and some animals.

Skill 3.10   Define antigen and antibody and recognize the antigen-antibody reaction

An **antigen** is any substance recognized by the body as foreign and thereby triggering an immune reaction. Bacteria, foreign cells, or foreign proteins may act as antigens.

An **antibody** is manufactured by the body and recognizes and latches onto antigens, hopefully destroying them. They also have recognition of foreign material versus the self. Memory of the invaders provides immunity upon further exposure.

Immunity is the body's ability to recognize and destroy an antigen before it causes harm.

There are two main responses made by the body after exposure to an antigen:

1. Humoral response - free antigens activate this response and B cells (a lymphocyte from bone marrow) give rise to plasma cells that secrete antibodies and memory cells that will recognize future exposures to the same antigen. The antibodies defend against extra cellular pathogens by binding to the antigen and making them an easy target for phagocytes to engulf and destroy. Antibodies are in a class of proteins called immunoglobulins. There are five major classes of immunoglobulins (Ig) involved in the humoral response: IgM, IgG, IgA, IgD, and IgE.

2. Cell mediated response - cells that have been infected activate T cells (a lymphocyte from the thymus). These activated T cells defend against pathogens in the cells or cancer cells by binding to the infected cell and destroying them along with the antigen. T cell receptors on the T helper cells recognize antigens bound to the body's own cells. T helper cells release IL-2, which stimulates other lymphocytes (cytotoxic T cells and B cells). Cytotoxic T cells kill infected host cells by recognizing specific antigens.

Vaccines are antigens given in very small amounts. They stimulate both humoral and cell mediated responses and memory cells recognize future exposure to the antigen so antibodies can be produced much faster.

Skill 3.11   Compare active and passive immunity, identifying the positive and negative effects of vaccines and inoculations

Immunity is the body's ability to recognize and destroy an antigen before it causes harm. Active immunity develops after recovery from an infectious disease (i.e. chicken pox) or after a vaccination (mumps, measles, rubella). In active immunity, ther person's own body produces the antibodies. Passive immunity may be passed from one individual to another and is not permanent. In passive immunity, antibodies from another person's body are used. A good example is the immunities passed from mother to nursing child. A baby's immune system is not well developed and the passive immunity they receive through nursing keeps them healthier.

Skill 3.12   Evaluate the roles of cell recognition (e.g., cell-to-cell signaling, autoimmune diseases, tissue rejection, cancer, pollen/stigma-style interaction) in normal and abnormal cell activity

Cell to cell communication or cell signaling is a complex system of communication that controls the basic cellular activities and coordinates cell actions.

The ability of cells to understand and correctly respond to their microenvironment is very important for their development, tissue repair, immunity and tissue homeostasis. Mistakes in the cellular communication and processing cause diseases like cancer, autoimmunity and diabetes. It is absolutely important that a better understanding of cell communication is important to treat diseases effectively and build artificial tissues.

The immune system attacks not only microbes but also cells that are not native of the host. This is the problem with skin grafts, organ transplantations, and blood transfusions. Antibodies to foreign blood and tissue types already exist in the body. If blood is transfused that is not compatible with the host, these antibodies destroy the new blood cells. There is a similar reaction when tissue and organs are transplanted.

The major histocompatibility complex (MHC) is responsible for the rejection of tissue and organ transplants. This complex is unique to each person. Cytotoxic T cells recognize the MHC on the transplanted tissue or organ as foreign and destroy these tissues. Various drugs are needed to suppress the immune system so this does not happen. The complication with this is that the patient is now more susceptible to infection.

Autoimmune disease occurs when the body's own immune system destroys its own cells. Lupus, Grave's disease, and rheumatoid arthritis are examples of autoimmune disease. There is no way to prevent autoimmune diseases. Immunodeficiency is a deficiency in either the humoral or cell mediated immune defenses. HIV is an example of an immunodeficiency disease.

Skill 3.13   Identify the effect of environmental factors on the biochemistry of living things (e.g., UV light effects on melanin and vitamin D production)

In this, we will discuss the effects of three factors i.e. UV, ozone and greenhouse effect on the biochemistry of living things.

1. UV

Ultraviolet light is an electromagnetic radiation with wavelength shorter than that of the visible light, but longer than soft x-rays. UV can be subdivided into near UV (380-200nm wavelength), far or vacuum UV (200-10nm, abbreviated as FUV or VUV), and extreme UV (1-31nm. abbreviated as EUV or XUV).

When talking about the effect of UV radiation on human health and the environment, the range of UV wavelengths is subdivided into UVA 380-315nm), also called Long Wave or "black light", UVB (315-280nm), also called Medium Wave, and UVC (<280nm), also called Short Wave or "germicidal".

In laser technology, the term deep ultraviolet or DUV refers to wavelengths below 300nm.

UVA, UVB and UVC can all damage collagen fibers accelerating the process of aging. In fact UVA is least harmful, but can cause aging of skin, DNA damage and possibly skin cancer. Because UVA light is of longer wavelength, it can penetrate deeper into the skin than UVB and is thought to be the main cause of wrinkles. UVB can also cause skin cancer. The radiation excites DNA molecules in the skin causing distortion of DNA helix, halted replication, gaps and misincorporation. These can lead to mutations, which can result in cancerous growths. Most of the suntan lotions available in the markets block only UVB, however, some are available that can block UVA as well.

High intensities of UVB light are dangerous to the eyes and proper protective gear has to be used. The UVB light can cause Welders' flash or photokeratitis or arc eye, cataracts, pterygium and pinguecula formation.

Beneficial effects of UV light: UV light induces the production of vitamin D in the skin. Some symptoms of vitamin D deficiency include increased illness, fatigue, and depression. Another effect of vitamin D deficiency is osteomalacia, which can result in bone pain, difficulty in weight bearing and sometimes fractures. UV radiation is used in the treatment of psoriasis and vitiligo. Both UVB and UVA radiation are used for this treatment along with psoralens. Most effective in psoriasis and vitiligo is UV light with wavelength of 311nm.

2. Greenhouse effect:

In simple terms, greenhouse effect means is a process by which an atmosphere warms a planet. This was first discovered by Joseph Fourier in 1824 and first investigated quantitatively by Svante Arrhenius in 1896. The name is from the similar effect, which greenhouses utilize to facilitate plant growth.

The earth receives lot of solar radiation. Just above the atmosphere, the direct solar radiation flux averages 1366 watts per square meter, after being distributed over the entire earth. This figure exceeds the power generated by human activities.

The solar radiation coming to earth is partly absorbed by chlorophyll and changed to the chemical bond energy of food, partly changed immediately to heat energy, and partly reflected as light.,. Because the atmosphere includes "greenhouse gases" such as methane and carbon dioxide, it is a good absorber of long wave thermal energy, and it effectively forms a one-way blanket over earth's surface.

Visible and near visible radiation from the sun easily gets through, but thermal radiation from the earth's surface can't easily get out. As a result, earth's surface warms up. This allows for the survival of life on our planet.

However, due to human activities, the concentration of carbon dioxide and methane in our atmosphere has increased dramatically in the last 150 years, primarily due to the burning of fossil fuels, deforestation, and the widespread use of cattle for food (which both decreases photosynthesis and increases methane). The result is a more powerful greenhouse effect, which results in global warming, which in turn results in multiple climate change effects since wind, ocean currents, and the water cycle are fueled by heat energy. In order to combat this anthropogenic ongoing climate change, multiple responses are needed including improving education for women and girls (which slows population growth), increasing the use of alternative (non fossil fuel) energy sources, decreasing meat consumption, planting forests, and improving public transportation and agricultural practices.

3. Ozone depletion:

The ozone layer or the ozonosphere layer is that part of the atmosphere which contains relatively high concentration of ozone. The ozone layer was discovered in 1913 by the French physicists Charles Fabry and Henri Buisson. The meteorologist G. M. B. Dobson, who developed a simple spectrophotometer that could be used to measure stratospheric oxygen from the ground, explored its properties in detail.

Although the concentration of ozone in the ozone layer is very small, it is vitally important to life because it absorbs biologically harmful UV radiation emitted by the sun. Depletion of the ozone layer could allow more of the UV radiation to reach earth's surface, causing increased genetic damage (discussed in UV section) to living things. The depletion of the ozone layer is mainly caused by chlorine and bromine radicals due to the release of large quantities of manmade organohalogen compounds, especially chlorofluorocarbons (CFCs) and bromofluorocarbons. Over the northern hemisphere, the ozone layer has been dropping by 4% per decade. Over approximately 5% of the earth's surface, around the north and south poles, much larger (but seasonal) declines have been seen and these are the ozone holes.

In the Montreal Protocol of 1987, the governments of many countries banned the use of CFCs and the production of CFCs is sharply declining. Compounds containing C-H bonds are designed to replace the function of CFCs since these compounds are more reactive and less likely to survive long enough in the atmosphere to reach the stratosphere where they could affect the ozone layer. Although the Montreal Protocol has resulted in some success in the recovery of the ozone layer, recent studies point to some further decreases in lower latitudes that are not yet understood, which may or may not be related to climate change events. More research is urgently needed to continue to protect life on Earth from harmful UV radiation.

Skill 3.14   Identify the roles of ATP and ADP in cellular processes

Adenosine triphosphate (ATP) is the energy currency or coin of the cell. ATP transfers energy from chemical bonds to energy absorbing (endergonic) reactions inside the cell. ATP consists of the nucleotide adenine (ribose sugar – 5 carbon sugar, adenine base and phosphate group) and two other phosphate groups. Energy is stored in the covalent bonds between phosphates, with the greatest amount of energy in the bond between the second and the third phosphate groups, called the pyrophosphate bond. Adenosine diphosphate (ADP) is formed when ATP is dephosphorylated by ATPases; ADP is converted back to ATP by ATP synthetases.

The chemical reaction for the synthesis of ATP is:

ADP + Pi + energy → ATP

Adenosine diphosphate + inorganic phosphate + energy produces Adenosine triphosphate.

ATP serves as the energy reservoir or storehouse for a cell. In a chemical reaction which breaks down ATP, energy is released for both anabolic and catabolic processes. The resulting ADP and phosphate can then be reused to form ATP again, recharged by the energy generated during other catabolic processes.

ATP → ADP + energy + Pi

Adenosine triphosphate → Adenosine diphosphate + energy + inorganic phosphate

In the light reaction of photosynthesis, the light energy excites the electrons in the molecule and a high energy electron is released from the chlorophyll and transfers some of its energy to a molecule of ADP, causing the ADP to bond to a third phosphate to form ATP. ATP is the energy storehouse, receiving the energy released from glucose in respiration. In the Calvin cycle, ATP is converted to ADP and releases energy for a series of reactions. Thus, ATP and ADP are the essential molecules for energy transfer in cells.

Skill 3.15   Compare chemosynthetic and photosynthetic processes and the roles of organisms using these processes in the ecosystem.

Chemosynthesis and photosynthesis are two chemically different pathways for the capturing of energy to make organic food molecules. What makes them similar is that they are both responsible for life. Chemosynthesis is carried out mainly by bacteria living in oceanic thermal vents. Because the conditions are so harsh, they are the only organisms able to survive there and must produce their own food source. They are in turn consumed by higher organisms and are thus, as the base of the food web, responsible for the sustenance of life. Plants, the main users of photosynthesis, are also at the bottom of their food chain. Plants (multi-cellular, eukaryotic organisms) obtain inorganic nutrients from the soil through their root systems and convert sunlight into chemical bond energy of organic molecules (food) through photosynthesis. Photosynthesis (reviewed in depth in section 3.6) is a second pathway by which organisms can produce their own food source. While both pathways result in the production of an energy/food source, the primary difference is that chemosynthesis is fueled by the oxidation of sulfites or ammonia and photosynthesis is fueled by sunlight. In chemosynthesis, bacteria manufacture carbohydrates and other organic molecules from the oxidization of sulfates or ammonia. The hydrogen they use comes from hydrogen sulfite, whereas the nitrogen comes from ammonia or nitrates.

Of the organisms that use chemosynthesis to produce biomass from 1-carbon molecules, two categories can be discerned. In rare instances sites may have hydrogen molecules (H_2) available, and the energy produced from the reaction between CO_2 and H_2 (leading to production of methane, CH_4) can be large enough to fuel the production of biomass. In most areas, energy for chemosynthesis is derived from reactions between O_2 and hydrogen sulfide/ ammonia. In this scenario, the chemosynthetic microorganisms are dependent on photosynthesis which occurs elsewhere to produce the O_2 that they require.

Hydrogen sulfide chemosynthesis - $CO_2 + O_2 + 4\{H_2S\} \rightarrow CH_2O + 4\{S\} + 3\{H_2O\}$

Photosynthesis- $6CO_2 + 6H_2O$ + energy (from sunlight) $\rightarrow C_6H_{12}O_6 + 6O$

Skill 3.16   Identify cell-to-cell communication in living things (e.g., electrical, molecular, ionic).

Cell to cell communication or cell signaling is a complex system of communication that controls the basic cellular activities and coordinates cell actions. The ability of cells to understand and correctly respond to their microenvironment is very important for their development, tissue repair, immunity and tissue homeostasis. Mistakes in the cellular communication and processing cause diseases like cancer, autoimmunity and diabetes. It is absolutely important that a better understanding of cell communication exists in order to treat diseases effectively and build artificial tissues.

Recent research gives us information on the underlying structure of cell signaling networks and how changes in these networks can affect the flow of information.

The first type of cell to cell signaling needs direct cell to cell contact. Some cells can form gap junctions that connect their cytoplasm to the one in the adjacent cells. In cardiac muscle, gap junctions allow for action potential propagation from the cardiac pace maker of the heart to spread and cause contraction of the heart.

The second type is the notch signaling mechanism in which two adjacent cells must make physical contact to communicate. This direct contact of cells facilitates a very precise control of cell differentiation during embryonic development.

Many cell signals are carried by molecules that are released by one cell and then move to make contact with another cell. Endocrine signals are called hormones. Produced by endocrine cells, hormones travel through blood to reach all parts of the body. Neurotransmitters can target only cells in the vicinity of the emitting cell. Some signaling molecules can function as both a hormone and a neurotransmitter. Norepinephrine can function as a hormone when released from the adrenal gland and can also be produced by neurons and function as a neurotransmitter in the brain.

Estrogen can be released by the ovary and function as a hormone or act locally via paracrine or autocrine signaling.

Cells receive information from their environment through proteins called receptors. Molecules that can activate or in some cases inhibit receptors can be classified as hormones, neurotransmitters, cytokines, growth factors and are generally termed as receptor ligands. The mechanism of ligand – receptor interactions are very important to cell signaling. Many receptors are cell surface proteins, whereas estrogen is a hydrophobic molecule.

Cell to cell communication is a very complex and dynamic process which is of utmost importance to cells.

Competency 4.0   **Knowledge of the interaction of Cell structure and function**

Skill 4.1   **Identify and analyze the major events in the development of the cell theory**

Cell theory states that all living organisms are made up of cells that are essentially the same chemically, are the fundamental unit of life, and arise from preexisting cells through cell division. Advances in microscopic technology initiated the development of cell theory.

The invention of the microscope in the 17th century allowed scientists to observe microscopic life and begin to identify microscopic structures. Athanasius Kircher and Antonie van Leeuwenhoek first observed maggots in decaying tissue and microorganisms (protozoa and other unicellular organisms) respectively. In 1665,

Robert Hooke provided the first description of the cell. Hooke labeled the microscopic units that made up a slice of cork as "cells".

Experiments by Lazzaro Spallanzani and Louis Pasteur in the 18th and 19th centuries clarified the distinction between living and non-living matter. Spallanzani and Pasteur showed that living organisms derive from other living organisms, thus disproving the theory of spontaneous generation, or life from non-life.

In 1838, botanist Matthias Schleiden theorized that cells and cellular products constitute all structural elements of plants. A year later zoologist Theodor Schwann reached the same conclusion about the structural elements of animals. The findings of Schleiden and Schwann represent the initial formulation of cell theory.

Cell theory continued to develop in the late 19th century with the identification of protoplasmic organelles. Finally, Walther Flemming observed the components of the nucleus and nucleolus. Through staining, Flemming observed chromosomes during cell division and first coined the term mitosis. Understanding the mechanism of mitosis gave support to the idea that all cells arise from other cells.

Skill 4.2 Distinguish between the major structural characteristics of prokaryotic and eukaryotic cells.

The cell is the basic unit of all living things. There are three types of cells. They are prokaryotes, eukaryotes, and archaea. Archaea have some similarities with prokaryotes, but are as distantly related to prokaryotes as prokaryotes are to eukaryotes.

Prokaryotes

Prokaryotes consist only of bacteria and cyanobacteria (formerly known as blue-green algae). The classification of prokaryotes is in the diagram below.

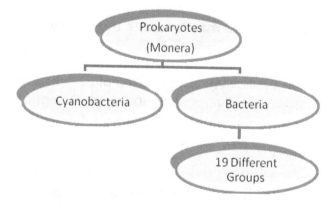

These cells have no defined nucleus or nuclear membrane. The DNA, RNA, and ribosomes float freely within the cell. The cytoplasm has a single chromosome condensed to form a **nucleoid**. Prokaryotes have a thick cell wall made up of amino sugars (glycoproteins). This is for protection, to give the cell shape, and to keep the cell from bursting. It is the **cell wall** of bacteria that is targeted by the antibiotic penicillin. Penicillin works by disrupting the cell wall, thus killing the cell.

The cell wall surrounds the **cell membrane** (plasma membrane). The cell membrane consists of a lipid bilayer that controls the passage of molecules in and out of the cell. Some prokaryotes have a capsule made of polysaccharides that surrounds the cell wall for extra protection from higher organisms.

Many bacterial cells have appendages used for movement called **flagella**. Some cells also have **pili**, which are a protein strand used for attachment of the bacteria. Pili may also be used for sexual conjugation (where the DNA from one bacterial cell is transferred to another bacterial cell).

Prokaryotes are the most numerous and widespread organisms on earth. Bacteria were most likely the first cells and date back in the fossil record to 3.5 billion years ago. Their ability to adapt to the environment allows them to thrive in a wide variety of habitats.

Eukaryotes

Eukaryotic cells are found in protists, fungi, plants, and animals. Most eukaryotic cells are larger than prokaryotic cells. They contain many organelles, which are membrane bound areas for specific functions. Their cytoplasm contains a cytoskeleton which provides a protein framework for the cell. The cytoplasm also supports the organelles and contains the ions and molecules necessary for cell function. The cytoplasm is contained by the plasma membrane. The plasma membrane allows molecules to pass in and out of the cell. The membrane can bud inward to engulf outside material in a process called endocytosis. Exocytosis is a secretory mechanism, the reverse of endocytosis.

The most significant differentiation between prokaryotes and eukaryotes is that the genetic material of eukaryotes is enclosed in a membrane-bound nucleus.

Skill 4.3   Relate the structure of cell organelles to their functions

The nucleus is the control center of the cell that contains all of the cell's genetic information. The chromosomes consist of chromatin, which is a complex of DNA and proteins. The chromosomes are tightly coiled to conserve space while providing a large surface area. The nucleus is the site of transcription of the DNA into RNA. The **nucleolus** is where ribosomes are made. There is at least one of these dark-staining bodies inside the nucleus of most eukaryotes. The nuclear envelope is two membranes separated by a narrow space. The envelope contains many pores that let RNA out of the nucleus.

Ribosomes are the site for protein synthesis. Ribosomes may be free floating in the cytoplasm or attached to the endoplasmic reticulum. There may be up to a half a million ribosomes in a cell, depending on how much protein is made by the cell.

The **endoplasmic reticulum** (ER) is folded and provides a large surface area. It is the "roadway" of the cell and allows for transport of materials through and out of the cell. There are two types of ER. Smooth endoplasmic reticulum contains no ribosomes on their surface. This is the site of lipid synthesis. Rough endoplasmic reticulum has ribosomes on their surface. They aid in the synthesis of proteins that are membrane bound or destined for secretion.

Many of the products made in the ER proceed on to the Golgi apparatus. The **Golgi apparatus** functions to sort, modify, and package molecules that are made in the other parts of the cell (like the ER). These molecules are either sent out of the cell or to other organelles within the cell. The Golgi apparatus is a stacked structure in order to increase the surface area.

Lysosomes are found mainly in animal cells. These contain digestive enzymes that break down food, substances not needed, viruses, damaged cell components and eventually the cell itself. It is believed that lysosomes are responsible for the aging process.

Mitochondria are large organelles that are the site of cellular respiration, where ATP is made to supply energy to the cell. Muscle cells have many mitochondria because they use a great deal of energy. Mitochondria have their own DNA, RNA, and ribosomes and are capable of reproducing by binary fission if there is a greater demand for additional energy. Mitochondria have two membranes: a smooth outer membrane and a folded inner membrane. The folds inside the mitochondria are called cristae. They provide a large surface area for cellular respiration to occur.

Plastids are found only in photosynthetic organisms. They are similar to the mitochondria due to the double membrane structure. They also have their own DNA, RNA, and ribosomes and can reproduce if the need for the increased capture of sunlight becomes necessary. There are several types of plastids. **Chloroplasts** are the site of photosynthesis. The stroma is the chloroplast's inner membrane space. The stoma encloses sacs called thylakoids that contain the photosynthetic pigment chlorophyll. The chlorophyll traps sunlight inside the thylakoid to generate ATP which is used in the stroma to produce carbohydrates and other products. The **chromoplasts** make and store yellow and orange pigments. They provide color to leaves, flowers, and fruits. The **amyloplasts** store starch and are used as a food reserve. They are abundant in roots like potatoes.

The Endosymbiotic Theory states that mitochondria and chloroplasts were once free living and possibly evolved from prokaryotic cells. At some point in our evolutionary history, they entered the eukaryotic cell and maintained a symbiotic relationship with the cell, with both the cell and organelle benefiting from the relationship. The fact that they both have their own DNA, RNA, ribosomes, and are capable of reproduction helps to confirm this theory.

Found in plant cells only, the **cell wall** is composed of cellulose and fibers. It is thick enough for support and protection, yet porous enough to allow water and dissolved substances to enter. **Vacuoles** are found mostly in plant cells. They hold stored food and pigments. Their large size allows them to fill with water in order to provide turgor pressure. Lack of turgor pressure causes a plant to wilt.

Below is a diagram of a generalized animal cell.

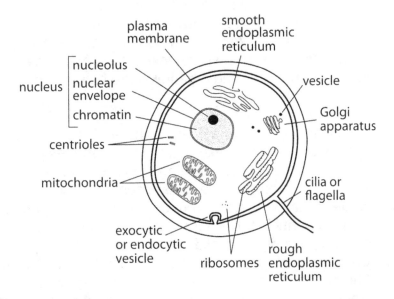

The **cytoskeleton**, found in both animal and plant cells, is composed of protein filaments attached to the plasma membrane and organelles. They provide a framework for the cell and aid in cell movement. They constantly change shape and move about. Three types of fibers make up the cytoskeleton:

1. **Microtubules** – the largest of the three, they make up cilia and flagella for locomotion. Some examples are sperm cells, cilia that line the fallopian tubes and tracheal cilia. Centrioles are also composed of microtubules. They aid in cell division to form the spindle fibers that pull the cell apart into two new cells. Centrioles are not found in the cells of higher plants.
2. **Intermediate filaments** – intermediate in size, they are smaller than microtubules but larger than microfilaments. They help the cell to keep its shape.
3. **Microfilaments** – smallest of the three, they are made of actin and small amounts of myosin (like in muscle tissue). They function in cell movement like cytoplasmic streaming, endocytosis, and ameboid movement. This structure pinches the two cells apart after cell division, forming two new cells.

Skill 4.4 Identify and evaluate the events of each phase of the cell cycle (G1, S, G2, M) and the regulatory mechanisms of the cycle

The purpose of cell division is to provide growth and repair in body (somatic) cells and to replenish or create sex cells for reproduction. There are two forms of cell division. **Mitosis** is the division of somatic cells and **meiosis** is the division of special cells to form gametes or sex cells (eggs and sperm).

Mitosis is divided into two parts: the **mitotic (M) phase** and **interphase**. In the mitotic phase, mitosis and cytokinesis divide the nucleus and cytoplasm, respectively. This phase is the shortest phase of the cell cycle. Interphase is the stage where the cell grows and copies the chromosomes in preparation for the mitotic phase. Interphase occurs in three stages of growth: **G1** (growth) period is when the cell is growing and metabolizing, the **S** period (synthesis) is where new DNA is being made and the **G2** phase (growth) is where new proteins and organelles are being made to prepare for cell division.

The mitotic phase is a continuum of change, although it is described as occurring in five stages: prophase, prometaphase, metaphase, anaphase, and telophase. During **prophase**, the cell proceeds through the following steps continuously, with no stopping. The chromatin condenses to become visible chromosomes; each chromosome has been duplicated so appears as two sister chromatids joined at the centromere. The nucleolus disappears and the nuclear membrane breaks apart. Mitotic spindles form that will eventually pull the pairs of chromatids apart. They are composed of microtubules. The cytoskeleton breaks down and the spindles are pushed to the poles or opposite ends of the cell by the action of centrioles. During **prometaphase**, the nuclear membrane fragments and allows the spindle microtubules to interact with the chromosomes. Kinetochore fibers attach to the chromosomes at the centromere region. (Sometimes prometaphase is grouped with metaphase). When the centrosomes are at opposite ends of the cell, the division is in **metaphase**. The centromeres of all the chromosomes are aligned with one another. During **anaphase**, the centromeres split in half and identical sister chromatids separate. The chromosomes are pulled to the poles of the cell, with identical sets at either end. The last stage of mitosis is **telophase**. Here, two nuclei form with a full set of DNA that is identical to the parent cell. The nucleoli become visible and the nuclear membrane reassembles. A cell plate is seen in plant cells, whereas a cleavage furrow is formed in animal cells. The cell is pinched into two cells. Cytokinesis, or division of the cytoplasm and organelles, occurs.

Mitosis

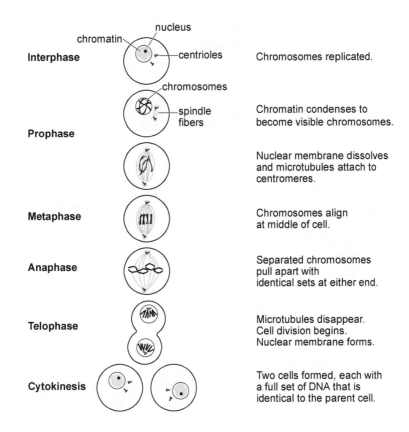

Interphase	Chromosomes replicated.
Prophase	Chromatin condenses to become visible chromosomes.
	Nuclear membrane dissolves and microtubules attach to centromeres.
Metaphase	Chromosomes align at middle of cell.
Anaphase	Separated chromosomes pull apart with identical sets at either end.
Telophase	Microtubules disappear. Cell division begins. Nuclear membrane forms.
Cytokinesis	Two cells formed, each with a full set of DNA that is identical to the parent cell.

Meiosis is similar to mitosis, but there are two consecutive cell divisions, meiosis I (when homologous pairs separate) and meiosis II (when chromatids separate), in order to reduce the chromosome number by one half. This way, when the sperm and egg join during fertilization, the haploid number is reached.

Meiosis

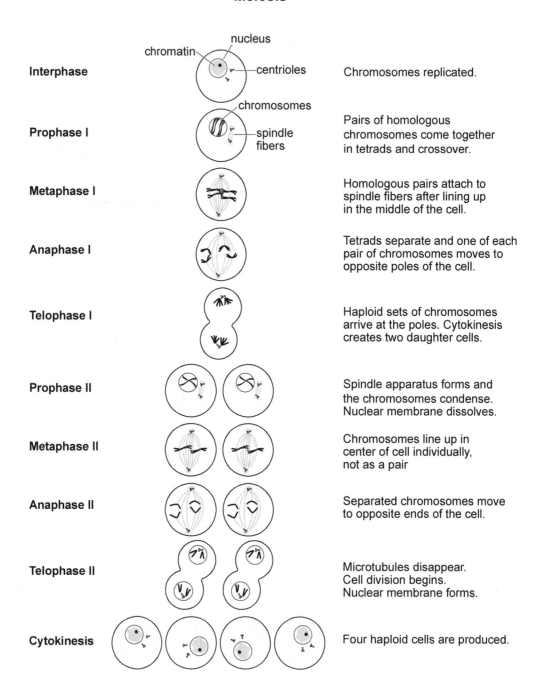

Interphase	Chromosomes replicated.
Prophase I	Pairs of homologous chromosomes come together in tetrads and crossover.
Metaphase I	Homologous pairs attach to spindle fibers after lining up in the middle of the cell.
Anaphase I	Tetrads separate and one of each pair of chromosomes moves to opposite poles of the cell.
Telophase I	Haploid sets of chromosomes arrive at the poles. Cytokinesis creates two daughter cells.
Prophase II	Spindle apparatus forms and the chromosomes condense. Nuclear membrane dissolves.
Metaphase II	Chromosomes line up in center of cell individually, not as a pair
Anaphase II	Separated chromosomes move to opposite ends of the cell.
Telophase II	Microtubules disappear. Cell division begins. Nuclear membrane forms.
Cytokinesis	Four haploid cells are produced.

Skill 4.5   Compare the mechanisms and results of nuclear division (karyokinesis) and cell division (cytokinesis) in plant and animal cells

The mechanisms and results of nuclear division (karyokinesis) are very similar in plant and animal cells. On the other hand, the mechanism of cytokinesis in plant and animal cells is significantly different because of the presence of a cell wall in plant cells. The main difference in nuclear division between plants and animals is the location within the organism where division occurs. Animal cells can divide throughout the organism, while plants' cell division is generally restricted to high growth areas called meristems (e.g. root tips).

The basic mechanics of replication and division of nuclear material is very similar in plant and animal cells. The slight differences arise from differences in cell structure (e.g. cell wall present in plants and not in animals) and preparations for cytokinesis. In prophase, the first phase of nuclear division, chromosomes condense and spindle formation begins. In prometaphase and metaphase, the nuclear envelope breaks down, chromosomes move to the metaphase plate, the spindle captures the chromosomes, and the chromosomes align on the spindle equator. In anaphase, chromosomes move toward cell poles. Finally, in telophase, the nuclear envelopes reform, the spindle breaks down, and chromosomes decondense.

The mechanism of plant and animal cell division (cytokinesis) differs greatly because of the presence of cell walls in plant cells. In animal cells, the first sign of cell division is a cleavage furrow, which begins as a shallow groove near the metaphase plate. A contractile ring, made of actin microfilaments, cleaves the cell, creating two separate and identical daughter cells with distinct nuclei and chromosomes. In plant cells, however, there is no cleavage furrow as vesicles from the Golgi apparatus gather in the middle of the replicating cell forming a cell plate. The vesicles contain cell wall materials that collect in the cell plate, and the cell plate enlarges and merges with the cell membrane to form a new cell wall between the still conjoined daughter cells.

Skill 4.6   Compare characteristics of the major taxa (domains/kingdoms), focusing on cellular characteristics

The traditional classification of living things is the five-kingdom system. The five kingdoms are Monera, Protista, Fungi, Plantae, and Animalia. The following is a comparison of the cellular characteristics of members of the five kingdoms.

Kingdom Monera

Members of the Kingdom Monera are single-celled, prokaryotic organisms. Like all prokaryotes, Monerans lack nuclei and other membrane bound organelles, but do contain circular chromosomes and ribosomes. Most Monerans possess a cell wall made of peptidoglycan, a combination of sugars and proteins.

Some Monerans also possess capsules and external motility devices (e.g. pili or flagella). The Kingdom Monera includes both eubacteria and archaebacteria. Though archaebacteria are structurally similar to eubacteria in many ways, there are key differences, like cell wall structure (archae lack peptidoglycan).

Kingdom Protista

Protists are eukaryotic, usually single-celled organisms (though some protists are multicellular). The Kingdom Protista is very diverse, containing members with characteristics of plants, animals, and fungi. All protists possess nuclei and some types of protists possess multiple nuclei. Most protists contain many mitochondria for energy production, and photosynthetic protists contain specialized structures called plastids where photosynthesis occurs. Motile protists possess external cilia or flagella. Finally, many protists have cell walls that do not contain cellulose.

Kingdom Fungi

Fungi are eukaryotic organisms that are mostly multicellular (single-celled yeast are the exception). Fungi possess cell walls composed of chitin. Fungal organelles are similar to animal organelles. Fungi are non-photosynthetic and possess neither chloroplasts nor plastids. Many fungal cells, like animal cells, possess centrioles. Fungi are also non-motile and release exoenzymes into the environment to dissolve food.

Kingdom Plantae

Plants are eukaryotic, multicellular, and have straight-edged cells. Plant cells possess rigid cell walls composed mostly of cellulose. Plant cells also contain chloroplasts and plastids for photosynthesis. Plant cells generally do not possess centrioles. Another distinguishing characteristic of plant cells is the presence of a large, central vacuole that occupies 50-90% of the cell interior. The vacuole stores acids, sugars, and wastes. Because of the presence of the vacuole, the cytoplasm is limited to a very small part of the cell.

Kingdom Animalia

Animals are eukaryotic, multicellular, and motile. Animal cells do not possess cell walls or plastids, but do possess a complex system of organelles. Most animal cells also possess centrioles, microtubule structures that play an important role in spindle formation during replication.

The three-domain system of classification, introduced by Carl Woese in 1990, emphasizes the separation of the two types of prokaryotes. The following is a comparison of the cellular characteristics of members of the three domains of living organisms: Eukarya, Bacteria, and Archaea.

Domain Eukarya

The Eukarya domain includes all members of the protist, fungi, plant, and animal kingdoms. Eukaryotic cells possess a membrane bound nucleus and other membranous organelles (e.g. mitochondria, Golgi, ribosomes). The chromosomes of Eukarya are linear and usually complexed with histones (protein spools). The cell membranes of eukaryotes consist of glycerol-ester lipids and sterols. The ribosomes of eukaryotes are 80 Svedburg (S) units in size. Finally, the cell walls of those eukaryotes that have them (i.e. plants, algae, fungi) are polysaccharide in nature.

Domain Bacteria

Prokaryotic members of the Kingdom Monera not classified as Archaea, are members of the Bacteria domain. Bacteria lack a defined nucleus and other membranous organelles. The ribosomes of bacteria measure 70 S units in size. The chromosome of Bacteria is usually a single, circular molecule that is not complexed with histones. The cell membranes of Bacteria lack sterols and consist of glycerol-ester lipids. Finally, most Bacteria possess a cell wall made of peptidoglycan.

Domain Archaea

Members of the Archaea domain are prokaryotic and similar to bacteria in most aspects of cell structure and metabolism. However, transcription and translation in Archaea are similar to the processes of eukaryotes, not bacteria. In addition, the cell membranes of Archaea consist of glycerol-ether lipids in contrast to the glycerol-ester lipids of eukaryotic and bacterial membranes. Finally, the cell walls of Archaea are not made of peptidoglycan, but consist of other polysaccharides, protein, and glycoprotein.

Skill 4.7   Evaluate the relationships between the structures and functions of cell membrane elements

The **cell membrane** (plasma membrane) consists of a lipid bilayer with protein molecules embedded in it.

All organisms contain cell membranes because they regulate the flow of materials into and out of the cell. The current model for the cell membrane is the Fluid Mosaic Model, which takes into account the ability of lipids and proteins to move and change places, giving the membrane fluidity.

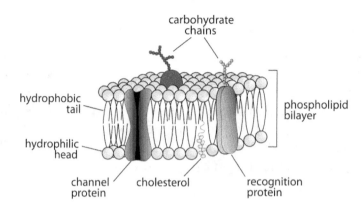

Cell membranes have the following characteristics:

They are made of phospholipids which have polar, charged heads with a phosphate group which is hydrophilic (water loving) and two nonpolar lipid tails which are hydrophobic (water fearing). This allows the membrane to orient itself with the polar heads facing the fluid inside and outside the cell and the hydrophobic lipid tails sandwiched in between. Hydrogen bonding holds the membrane together. Each individual phospholipid is called a micelle.

They contain proteins embedded inside (integral proteins), proteins that stretch through (transmembrane proteins), and proteins on the surface (peripheral proteins). These proteins may act as channels for transport, may contain enzymes, may act as receptor sites, may act to stick cells together, or may attach to the cytoskeleton to give the cell shape

They contain cholesterol, which alters the fluidity of the membrane.

They contain oligosaccharides (small carbohydrate polymers) on the outside of the membrane. These act as markers that help distinguish one cell from another.

They contain receptors made of glycoproteins that can attach to certain molecules, like hormones.

Skill 4.8   Identify and compare active and passive transport mechanisms.

Cell transport is necessary to maintain homeostasis, or balance of the cell with its external environment. Cell membranes are selectively permeable, which is the key to transport. Not all molecules may pass through easily. Some molecules require energy or carrier molecules and may only cross when needed.

Passive transport does not require energy and moves the material with the concentration gradient (high to low). Small molecules may pass through the membrane in this manner. Two examples of passive transport include diffusion and osmosis. **Diffusion** is the natural tendency of molecules, which are in motion, to move from areas of high concentration to areas of low concentration. **Osmosis** is simply the diffusion of water across a semi-permeable membrane. Osmosis may cause cells to swell or shrink, depending on the internal and external environments. The following terms are used in relation of the cell to the environment.

Isotonic - water concentration is equal inside and outside the cell. Movement in either direction is basically equal, so net change in concentration is zero.

Hypertonic - "hyper" refers to the amount of dissolved particles. The more particles are in a solution, the lower its water concentration. Therefore, when a cell is hypertonic to its environment, there is more water outside the cell than inside. Water will move into the cell and the cell will swell. If the environment is hypertonic to the cell, there is more water inside the cell. Water will move out of the cell and the cell will shrink.

Hypotonic - "hypo" again refers to the amount of dissolved particles. The less particles in a solution, the higher its water concentration. When a cell is hypotonic to its environment, there is more water inside the cell than outside. Water will move out of the cell and the cell will shrink. If the environment is hypotonic to the cell, there is more water outside the cell than inside. Water will move into the cell and the cell will swell.

The **facilitated diffusion** mechanism does not require energy, but does require a carrier protein. An example would be insulin, which is needed to carry glucose into the cell.

Active transport requires energy. The energy for this process comes from either ATP or an electrical charge difference. Active transport may move materials either with or against a concentration gradient. Some examples of active transport are:

- Calcium pumps - actively pump calcium outside of the cell and are important in nerve and muscle transmission.
- Stomach acid pump - exports hydrogen ions to lower the pH of the stomach and increase acidity.
- Sodium-Potassium pump - maintains an electrical difference across the cell. This is useful in restoring ion balance so nerves can continue to function. It exchanges sodium ions for potassium ions across the plasma membrane in animal cells.

Active transport involves a membrane potential which is a charge on the membrane. The charge works like a magnet and may cause transport proteins to alter their shape, thus allowing substances in or out of the cell.

Competency 5.0   **Knowledge of Genetic Principles and Practices**

Skill 5.1   **Evaluate the relationships between the structure and function of DNA**

Nucleic acids consist of DNA (deoxyribonucleic acid) and RNA (ribonucleic acid). Nucleic acids contain the instructions for the amino acid sequence of proteins and the instructions for replicating. The monomer of nucleic acids is called a nucleotide. A nucleotide consists of a 5 carbon sugar, (deoxyribose in DNA, ribose in RNA), a phosphate group, and a nitrogenous base. The base sequence codes for the instructions. There are five bases: adenine, thymine, cytosine, guanine, and uracil. Uracil is found only in RNA and replaces the thymine. A summary of nucleic acid structure can be seen in the table below:

	SUGAR	PHOSPHATE	BASES
DNA	Deoxyribose	Present	adenine, **thymine**, cytosine, guanine
RNA	Ribose	Present	adenine, **uracil**, cytosine, guanine

Due to the molecular structure, adenine will always pair with thymine in DNA or uracil in RNA. Cytosine always pairs with guanine in both DNA and RNA. This allows for the symmetry of the DNA molecule seen below.

Adenine and thymine (or uracil) are linked by two covalent bonds and cytosine and guanine are linked by three covalent bonds. The guanine and cytosine bonds are harder to break apart than thymine (uracil) and adenine because of the greater number of these bonds. The DNA molecule is called a double helix due to its twisted ladder shape.

Skill 5.2 Identify and sequence the principal events in DNA replication

DNA replicates semi-conservatively. This means the two original strands are conserved and serve as a template for the new strand.

In DNA replication, the first step is to separate the two strands. As they separate, they need to unwind the supercoils to reduce tension. An enzyme called **helicase** unwinds the DNA as the replication fork proceeds and **topoisomerases** relieve the tension by nicking one strand and letting the supercoil relax. Once the strands have been separated, they need to be stabilized.

Single stranded binding proteins (SSBs) bind to the single strands until the DNA is replicated.

DNA Replication

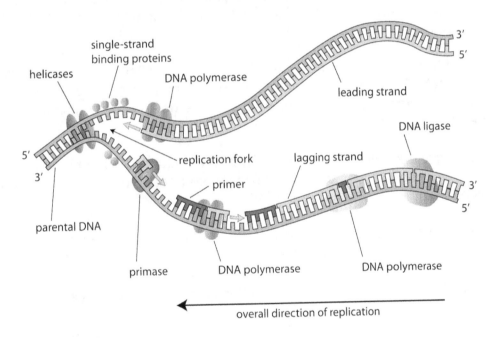

single-strand binding proteins

helicases

DNA polymerase

leading strand

DNA ligase

replication fork

lagging strand

parental DNA

primer

primase

DNA polymerase

DNA polymerase

overall direction of replication

Skill 5.3 Identify and sequence the principal events of protein synthesis

Proteins are synthesized through the processes of transcription and translation. Three major classes of RNA are needed to carry out these processes. The first is **messenger RNA (mRNA)**, which contains information for translation. **Ribosomal RNA (rRNA)** is a structural component of the ribosome and **transfer RNA (tRNA)** carries amino acids to the ribosome for protein synthesis.

Transcription is similar in prokaryotes and eukaryotes. During transcription, the DNA molecule is copied into an RNA molecule (mRNA). Thus the word transcription suggests that the language of nucleic acids has not changed, but the form of that language has changed. Transcription occurs through the steps of initiation, elongation, and termination. Transcription also occurs for rRNA and tRNA, but the focus here is on mRNA.

Initiation begins at the promoter of the double stranded DNA molecule. The promoter is a specific region of DNA that directs the **RNA polymerase** to bind to the DNA. The double stranded DNA opens up and RNA polymerase begins transcription in the 5' → 3' direction by pairing ribonucleotides to the deoxyribonucleotides as follows to get a complementary mRNA segment:

Deoxyribonucleotide		Ribonucleotide
A	→	U
G	→	C

Elongation is the synthesis on the mRNA strand in the 5' → 3' direction. The new mRNA rapidly separates from the DNA template and the complementary DNA strands pair together again.

Termination of transcription occurs at the end of a gene. Cleavage occurs at specific sites on the mRNA. This process is aided by termination factors. In eukaryotes, mRNA goes through **posttranscriptional processing** before going on to translation. There are three basic steps of processing:

1. 5' capping is attaching a base with a methyl attached to it that protects 5' end from degradation and serves as the site where ribosome binds to mRNA for translation.

2. 3' polyadenylation is when about 100-300 adenines are added to the free 3' end of mRNA resulting in a poly-A-tail.

3. Introns (non-coding) are removed and the coding exons are spliced together to form the mature mRNA.

Translation is the process in which the mRNA sequence is translated into a "new language" of proteins: amino acids will be put together in a particular order, based on the nucleic acid code, to become a polypeptide. The mRNA sequence determines the amino acid sequence of a protein by following a pattern called the genetic code. The **genetic code** consists of triplet nucleotide combinations called **triplet codons.**. There are 20 amino acids mRNA codes for. Amino acids are the building blocks of protein. They are attached together by peptide bonds to form a polypeptide chain. There are 64 triplet combinations called codons. Three codons are termination codons and the remaining 61 code for amino acids.

Ribosomes are the site of translation. They contain rRNA and many proteins. Translation occurs in three steps: initiation, elongation, and termination. Initiation occurs when the methylated tRNA binds to the ribosome to form a complex. This complex then binds to the 5' cap of the mRNA. In elongation, tRNAs carry the amino acid to the ribosome and place it in order according to the mRNA sequence. tRNA is very specific – it only accepts one of the 20 amino acids that corresponds to itsanticodon. The anticodon is complementary to the codon. For example, using the codon sequence below:

the mRNA reads   A U G / G A G / C A U / G C U
the anticodons are   U A C / C U C / G U A / C G A

Termination occurs when the ribosome reaches any one of the stop codons: UAA, UAG, or UGA. The newly formed polypeptide then undergoes posttranslational modification to alter or remove portions of the polypeptide.

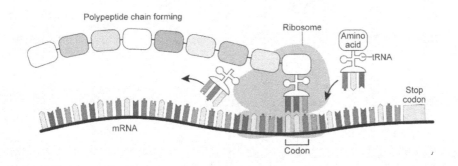

Skill 5.4   Distinguish between the various functions of DNA and RNA

The primary function of DNA is storage of genetic information. DNA also directs the production of proteins and RNA, thus controlling cellular biochemistry. Finally, replicated DNA conserves and transmits genetic information to future generations.

In contrast to DNA, there are three main types of RNA with distinct functions. Messenger RNA (mRNA) carries information from DNA during transcription to ribosomes where the information is translated into the gene product, RNA or protein. Transfer RNA (tRNA) is the adapter molecule of protein synthesis. Each tRNA contains an amino acid and a three-base anticodon. The anticodon recognizes a three-base region of the mRNA through complimentary base pairing and delivers its amino acid to the growing polypeptide chain. Finally, ribosomal RNA (rRNA) is a structural and functional component of cellular ribosomes.

Skill 5.5   Distinguish between the regulatory systems for prokaryotic and eukaryotic protein synthesis

Controlling the rate of transcriptional activity is the main method of prokaryotic protein synthesis regulation. Prokaryotes generally regulate gene expression and protein synthesis with simple operon systems that tie protein synthesis to metabolic activity.

Prokaryotic DNA molecules possess sequence elements called promoters that RNA polymerase recognizes as start sites for transcription. Repressor proteins prevent binding of RNA polymerase to the promoter regions.

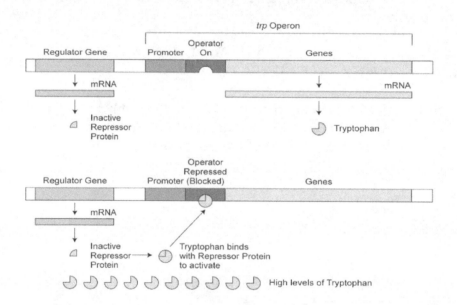

Prokaryotic cells express groups of genes related to the utilization of specific energy sources (e.g. lactose) only when there are large amounts of the energy source present. For example, the high presence of lactose prevents the *lac* operon repressor from binding and promotes expression of lactose metabolism genes. Prokaryotic cells regulate expression of genes responsible for synthesis of small biomolecules through transcription repression and attenuation, termination of transcription before product formation. For example, the genes that produce the proteins responsible for tryptophan, an amino acid, actively express when cellular tryptophan levels are low. High levels of tryptophan promote repressor binding and attenuation of transcribing proteins.

Eukaryotic cells, in contrast to prokaryotic cells, have many methods of protein synthesis regulation at different points in the process.

Chromatin Structure: Because eukaryotic DNA is compacted into chromatin and often complexed with histones, cells can restrict the access of RNA polymerase and other transcription factors to the DNA.

Transcription Initiation: As in prokaryotes, this is the most important method of regulation. As expected, the controls in eukaryotes are more complex. Various promoter and enhancer regions of the DNA interact with multiple activator and inhibitor proteins to control gene expression.

Transcript Processing: Because eukaryotic cells have a distinct nucleus that segregates transcription and translation, transcribed mRNAs are processed prior to translation.

Protein elements modify and edit the mRNA, removing non-coding regions and splicing the mRNA to promote the expression of different genes.

Translation Initiation: Finally, eukaryotic cells regulate protein synthesis by controlling the rate of translation of mRNA into proteins. The amount and type of tRNAs present in a cell can affect the overall rate of translation.

Skill 5.6   Evaluate the appropriate application of DNA manipulation techniques (e.g. gene splicing, recombinant DNA, gene identification, PCR technique)

The manipulation of DNA has many applications including gene splicing, recombinant DNA technology, gene identification, and PCR technology.

Gene splicing is the addition of base pairs from another organism to plasmid DNA. Chemicals called restriction enzymes excise a portion of the plasmid DNA at specific and unique nucleotide sequences. The resulting DNA strand has two single stranded ends. Incubation of the cut DNA with a foreign genetic sequence and the enzyme ligase results in the creation of a spliced gene, also called **recombinant DNA**. Gene splicing and recombinant DNA technology has many medicinal applications. For example, plasmids spliced with the human gene for insulin production can be inserted into bacteria. As the bacteria grow and multiply, they produce large quantities of insulin used for treatment of diabetes. In addition, gene therapy, the splicing of genes into an individual's cells and tissues, is a promising treatment for diseases such as arthritis. Other applications of recombinant DNA and gene splicing include genetic modification of plants (e.g. introduction of insect or drought resistance genes), vaccine development, and pharmaceutical development.

DNA sequencing, determining the order of bases, allows for identification of genes. Computer programs and algorithms help predict the location of genes by examining characteristics of the DNA base sequences. In addition, examination of mRNA harvested from cells allows the identification of both DNA sequences that code for the mRNA and amino acid sequences of the proteins produced from the mRNA.

The **polymerase chain reaction (PCR)** is a technique of molecular biology for rapidly replicating DNA without using a living organism as a vector. PCR allows the production of large amounts of DNA from a small sample. Applications of PCR include the identification of genetic fingerprints, identification of hereditary disease, gene cloning, and paternity testing.

Skill 5.7   **Predict the effects of environmental and other influences on gene structure and expression (e.g. viruses, oncogenes, carcinogenic agents, mutagenic agents)**

Environmental factors can influence the structure and expression of genes. For instance, viruses can insert their DNA into the host's genome changing the composition of the host DNA. In addition, mutagenic agents found in the environment cause mutations in DNA and carcinogenic agents promote cancer, often by causing DNA mutations.

Many viruses can insert their DNA into the host genome causing mutations.

Many times viral insertion of DNA does not harm the host DNA because of the location of the insertion. Some insertions, however, can have grave consequences for the host. Oncogenes are genes that increase the malignancy of tumor cells. Some viruses carry oncogenes that, when inserted into the host genome, become active and promote cancerous growth. In addition, insertion of other viral DNA into the host genome can stimulate expression of host proto-oncogenes, genes that normally promote cell division. For example, insertion of a strong viral promoter in front of a host proto-oncogene may stimulate expression of the gene and lead to uncontrolled cell growth (i.e. cancer).

In addition to viruses, physical and chemical agents found in the environment can damage gene structure. Mutagenic agents cause mutations in DNA. Examples of mutagenic agents are x-rays, UV light, and ethidium bromide. Carcinogenic agents are any substances that promote cancer. Carcinogens are often, but not always, mutagens. Examples of agents carcinogenic to humans are asbestos, UV light, x-rays, and benzene.

Skill 5.8   **Analyze the processes and products of meiosis (e.g., gametogenesis in male and female vertebrates; plant, animal and fungi meiosis) in representative examples from various kingdoms**

Meiosis is preceded by an interphase during which the chromosome replicates. The steps of meiosis are as follows:

1. **Prophase I** – the replicated chromosomes condense and pair with homologues in a process called synapsis. This forms a tetrad. Crossing over, the exchange of genetic material between homologues to further increase diversity, occurs during prophase I.

2. **Metaphase I** – the homologous pairs attach to spindle fibers after lining up in the middle of the cell.

3. **Anaphase I** – the sister chromatids remain joined, while homologous pairs of chromosomes separate and move to the poles of the cell.

4. **Telophase I** – the homologous chromosome pairs continue to separate. Each pole now has a haploid chromosome set. Telophase I occurs simultaneously with cytokinesis. In animal cells, cleavage furrows form and cell plate appear in plant cells.

5. **Prophase II** – a spindle apparatus forms and the chromosomes condense.

6. **Metaphase II** – sister chromatids line up in center of cell. The centromeres divide and the sister chromatids begin to separate.

7. **Anaphase II** – the separated chromosomes move to opposite ends of the cell.

8. **Telophase II** – cytokinesis occurs, resulting in four haploid daughter cells.

Also refer to the meiosis diagram on page 62.

Skill 5.9 **Differentiate between classical laws of inheritance, their relationship to chromosomes, and related terminology**

Gregor Mendel is recognized as the father of genetics. His work in the late 1800s is the basis of our knowledge of genetics. Although unaware of the presence of DNA or genes, Mendel realized there were factors (now known as **genes**) that were transferred from parents to their offspring. Mendel worked with pea plants and fertilized the plants himself, keeping track of subsequent generations which led to the Mendelian laws of genetics. Mendel found that two "factors" governed each trait, one from each parent. Traits or characteristics came in several forms, known as **alleles**. For example, the trait of flower color had white alleles (*pp*) and purple alleles (*PP*). Mendel formed two laws: the law of segregation and the law of independent assortment.

The **law of segregation** states that only one of the two possible alleles from each parent is passed on to the offspring. If the two alleles differ, then one is fully expressed in the organism's appearance (the dominant allele) and the other has no noticeable effect on appearance (the recessive allele). The two alleles for each trait segregate into different gametes. A Punnet square can be used to show the law of segregation. In a Punnet square, one parent's genes are put at the top of the box and the other parent's on the side. Genes combine in the squares just like numbers are added in addition tables. This Punnet square shows the result of the cross of two F_1 hybrids.

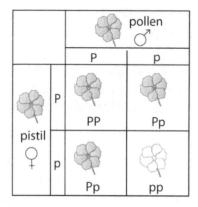

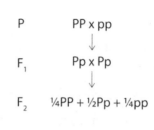

$$P \quad PP \times pp$$
$$\downarrow$$
$$F_1 \quad Pp \times Pp$$
$$\downarrow$$
$$F_2 \quad \tfrac{1}{4}PP + \tfrac{1}{2}Pp + \tfrac{1}{4}pp$$

This cross results in a 1:2:1 genotypic ratio of F_2 offspring. Here, the *P* is the dominant allele and the *p* is the recessive allele. The F_1 cross produces a phenotypic ratio of 3:1, meaning approximately three out of four offspring will have the dominant allele expressed (two *PP* and *Pp*) and one out four offspring will have the recessive allele expressed (*pp*).

Some other important terms to know:

> **Homozygous** – having a pair of identical alleles. For example, *PP* and *pp* are homozygous pairs.

> **Heterozygous** – having two different alleles. For example, *Pp* is a heterozygous pair.

> **Phenotype** – the organism's physical appearance.

> **Genotype** – the organism's genetic makeup. For example, *PP* and *Pp* have the same phenotype (purple in color), but different genotypes.

The **law of independent assortment** states that different pairs of alleles sort independently of each other. This law applies when more than one character trait is followed in a given cross. In a dihybrid cross, two characters are being explored.

Two of the seven characters Mendel studied were seed shape and color. Yellow is the dominant seed color (*Y*) and green is the recessive color (*y*). The dominant seed shape is round (*R*) and the recessive shape is wrinkled (*r*). A cross between a plant with yellow round seeds (*YYRR*) and a plant with green wrinkled seeds (*yyrr*) produces an F_1 generation with the genotype *YyRr*. The production of F_2 offspring results in a 9:3:3:1 phenotypic ratio.

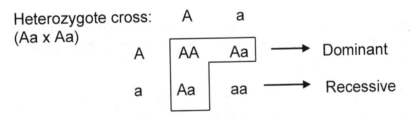

		pollen ♂			
		YR	Yr	yR	yr
pistil ♀	YR	◯ YYRR	◯ YYRr	◯ YyRR	◯ YyRr
	Yr	◯ YYRr	☁ YYrr	◯ YyRr	☁ Yyrr
	yR	◯ YyRR	◯ YyRr	◯ yyRR	◯ yyRr
	yr	◯ YyRr	☁ Yyrr	◯ yyRr	☁ yyrr

P   YYRR x yyrr
        ↓
F₁   YyRr
        ↓
F₂   1 - YYRR
      2 - YYRr } ◯ 9 yellow round
      2 - YyRR
      4 - YyRr

      1 - yyRR } ◯ 3 green round
      2 - yyRr

      1 - YYrr } ☁ 3 yellow wrinkled
      2 - Yyrr

      1 - yyrr } ☁ 1 green wrinkled

Skill 5.10   Analyze applications of probability and chi-square analysis in genetics

Scientists use probabilities to express genetic ratios. The genetic ratio of a given fertilization event represents the probability that a zygote will possess a given trait. For example, a monohybrid cross of two heterozygotes (Aa) for the dominant trait A yields a phenotypic ratio of 3:1 (three dominants to one recessive).

Heterozygote cross: A   a
(Aa x Aa)

A | AA Aa | ⟶ Dominant
a | Aa aa | ⟶ Recessive

Genetic ratios are merely hypothetical predictions based on several assumptions about the fertilization event. The assumptions are: (1) each allele is dominant or recessive, (2) alleles segregate (each parent provides only one allele), (3) independent assortment occurs (alleles for each trait are passed independently of each other), and (4) fertilization is random.

Because genetic ratios are hypothetical probabilities, scientists use a statistical test called chi-square (χ^2) analysis to evaluate observed deviation from the expected ratios in experimental genetic crosses. The χ^2 test reduces the observed deviation from each component of the expected ratio and the sample size to a single numerical value. The equation for calculating χ^2 is

$$\chi^2 = \Sigma \frac{(o-e)^2}{e}$$

In this equation,

> o = observed value of each phenotype
> e = expected value of each phenotype
> Σ = summation of the calculated values for each phenotype

Because $(o - e)$ is the deviation (d) in each case, the formula reduces to

$$\chi^2 = \frac{d^2}{e}$$

The following table shows the results of an experimental monohybrid cross with the χ^2 calculations.

Table 1. Chi-square analysis of monohybrid cross					
Ratio	Observed (o)	Expected (e)	Deviation ($d = o - e$)	d^2	d^2/e
3/4	720	750	720 − 750 = - 30	900	1.2
1/4	280	250	280 − 250 = + 30	900	3.6
Total = 1000 samples			χ^2 = 1.2 + 3.6 = 4.8		

The χ^2 value allows us to estimate the probability that the observed deviation is attributable strictly to chance. Because calculating the probability (p) from χ^2 is complex, we determine p from a graph or table of values.

In the table of p values (below) df is degrees of freedom, $n - 1$ where n is the number of different phenotypes. As the number of possible phenotypes or categories increases, the amount of expected deviation increases.

Table 2. χ^2 to p value conversion						
Probability (p)						
df	0.90	0.50	0.20	0.05	0.01	0.001
1	0.02	0.46	1.64	3.84	6.64	10.83
2	0.21	1.39	3.22	5.99	9.21	13.82
3	0.58	2.37	4.64	7.82	11.35	16.27
χ^2 values						

The p value represents the probability that the deviation in an experimental cross is attributable to chance. For example, the χ^2 value calculated in Table 1 of the hypothetical monohybrid cross is 4.8. There is one degree of freedom. From Table 2, the p value is approximately 0.035. Thus, the probability that the deviation in the observed results of the monohybrid cross is due to chance is 3.5%. The standard or critical value of p is 0.05. In other words, if there is less than a 5 percent possibility that the deviation is due to chance, we reject the hypothesis that the experiment was a heterozygote cross where the assumptions used to formulate the generic ratios are true. If the value of p is greater than 0.05, we accept the hypothesis that the experiment was a heterozygotic cross and the genetic assumptions were satisfied. Thus, in the above example, the experimental cross did not satisfy the assumptions. There are several possible explanations for experimental crosses that produce p values less than 0.05. For example, the parent organisms may not be heterozygotes, the trait may be linked to other genes, or the viability of certain genotypes may not be equal.

Skill 5.11   **Analyze various patterns of inheritance (e.g., sex-linked, sex-influenced, sex-limited, incomplete dominance, autosomal linkage, multiple alleles, polygenic inheritance)**

The same techniques of pedigree analysis apply when tracing inherited disorders. Thousands of genetic disorders are the result of inheriting a recessive trait. These disorders range from nonlethal traits (such as albinism) to life-threatening (such as cystic fibrosis).

Most people with recessive disorders are born to parents with normal phenotypes. The mating of heterozygous parents would result in an offspring genotypic ratio of 1:2:1; thus approximately 1 out of 4 offspring would express this recessive trait, with each offspring having a 25% chance of exhibiting the trait. That is, having one child with the trait does not decrease the chances that the next child will also have the trait. The heterozygous parents are called carriers because they do not express the trait phenotypically but pass the trait on to their offspring.

Lethal dominant alleles are much less common than lethal recessives. This is because lethal dominant alleles are not masked in heterozygotes.

Mutations in a gene of the sperm or egg can result in a lethal dominant allele, usually killing the developing offspring.

Sex linked traits – These are carried on the X chromosome. The Y chromosome found only in males (XY) is small and carries very little genetic information, whereas the X chromosome found in females (XX) carries more.. Since men have no second X chromosome to cover up a recessive gene, the recessive trait is expressed more often in men. Women need the recessive gene on both X chromosomes to show the trait. Examples of sex linked traits include hemophilia and color-blindness.

Sex influenced traits - traits are influenced by the sex hormones. Male pattern baldness is an example of a sex influenced trait. Testosterone influences the expression of the gene. Mostly men lose their hair due to this.

Nondisjunction - during meiosis, chromosomes fail to separate properly. One sex cell may get both chromosomes and another may get none. Depending on the chromosomes involved this may or may not be serious. Offspring end up with either an extra chromosome or are missing one. An example of nondisjunction is Down Syndrome, where three #21 chromosomes are present.

Skill 5.12 Identify the causes of genetic disorders (e.g., point mutation, nondisjunction, translocation, deletion, insertion, inversion, duplication)

Inheritable changes in DNA are called mutations. **Mutations** may be errors in replication or a spontaneous rearrangement of one or more segments by factors like radioactivity, drugs, or chemicals. The severity of the change is not as critical as where the change occurs. DNA contains large segments of non-coding areas called introns. The important coding areas are called exons. If an error occurs on an intron, there is no effect. If the error occurs on an exon, it may be minor to lethal depending on the severity of the mistake. Mutations may occur on somatic or sex cells. Usually the mutations on sex cells are more significant since they contain the basis of all information for the developing offspring, and they are the mutations that can be passed on. But mutations are not always bad. They are the basis of evolution and if they make a more favorable variation that enhances the organism's survival, then they are beneficial. But mutations may also lead to abnormalities and birth defects and even death. There are several types of mutations.

A **point mutation** is a mutation involving a single nucleotide or a few adjacent nucleotides. Let's suppose a normal sequence was as follows:

Normal	A B C D E F	
Duplication	A B **C C** D E F	one nucleotide is repeated
Inversion	A **E D C B** F	a segment of the sequence is flipped
Deletion	A B C E F	a nucleotide is left out (D is lost)
Insertion	A B C **R S** D E F	nucleotides are inserted or translocated
Breakage	A B C	a piece is lost (DEF is lost)

Deletion and insertion mutations that shift the reading frame are **frame shift mutations**. A **silent mutation** makes no change in the amino acid sequence, therefore it does not alter the protein function. A **missense mutation** results in an alteration in the amino acid sequence.

Skill 5.13   Identify the effect of a mutation in a DNA sequence on the products of protein synthesis

A mutation's effect on protein function depends on which amino acid is involved and how many are involved. The structure of a protein usually determines its function. A mutation that does not alter the structure will probably have little or no effect on the protein's function. However, a mutation that does alter the structure of a protein can severely affect protein activity is called **loss-of-function mutation**. Sickle-cell anemia and cystic fibrosis are examples of loss-of-function mutations.

Competency 6.0   Knowledge of the structural and functional diversity of Viruses and Prokaryotic organisms

Skill 6.1   Distinguish the structure and function of viruses and prokaryotic organisms

Microbiology includes the study of monera, protists and viruses. Although **viruses** are not classified as living things, they greatly affect other living things by disrupting cell activity. They are considered to be obligate parasites because they rely on the host for their own reproduction. Viruses are composed of a protein coat and a nucleic acid, either DNA or RNA. A bacteriophage is a virus that infects a bacterium. Animal viruses are classified by the type of nucleic acid, presence of RNA replicase, and presence of a protein coat.

There are two types of viral reproductive cycles:

1. **Lytic cycle** - the virus enters the host cell and makes copies of its nucleic acids and protein coats and reassembles. It then lyses or breaks out of the host cell and infects other nearby cells, repeating the process.
2. **Lysogenic cycle** - the virus may remain dormant within the cells until something initiates it to break out of the cell. Herpes is an example of a lysogenic virus.

Archaebacteria and eubacteria are the two main branches of prokaryotic (moneran) evolution. Archaebacteria evolved from the earliest cells. Most achaebacteria inhabit extreme environments. There are three main groups of archaebacteria. Methanogens are strict anaerobes, extreme halophiles live in high salt concentrations, and extreme thermophiles live in hot temperatures (hot springs).

Most prokaryotes fall into the eubacteria (bacteria) domain. **Bacteria** are divided according to their morphology (shape). Bacilli are rod shaped bacteria, cocci are round bacteria and spirilli are spiral shaped.

The Gram stain is a procedure used to differentiate the cellular make-up of bacteria. Gram positive bacteria have simple cell walls consisting of large amounts of peptidoglycan. These bacteria pick up the stain, revealing a purple color when observed under the microscope. Gram negative bacteria have a more complex cell wall consisting of less peptidoglycan, but have large amounts of lipopolysaccharides. The lipopolysaccharides resist the stain, revealing a pink color when observed under the microscope. Because of the lipopolysaccharide cell wall, Gram negative bacteria tend to be more toxic and are more resistant to antibiotics and host defense mechanisms.

Bacteria reproduce by binary fission. This asexual process is simply dividing the bacterium in half. All new organisms are exact clones of the parent. Some bacteria have a sticky capsule that protects the cell wall and is also used for adhesion to surfaces. Pili are surface appendages for adhesion to other cells.

Bacteria locomotion is via flagella or taxis. Taxis is the movement towards or away from a stimulus. The methods for obtaining nutrition are: for photosynthetic organisms or producers- the conversion of sunlight to chemical energy, consumers or heterotrophs- consuming other living organisms, and saprophytes are consumers that live off dead or decaying material.

In comparison, archaebacteria contain no peptidoglycan in the cell wall, they are not inhibited by antibiotics, they have several kinds of RNA polymerase, and they do not have a nuclear envelope. Eubacteria (bacteria) have peptidoglycan in the cell wall, they are susceptible to antibiotics, they have one kind of RNA polymerase, and they have no nuclear envelope.

Skill 6.2　　Identify the effects of viruses (e.g., HIV, influenza, measles, TMV, feline leukemia, genital warts, some human cancers) on organisms

Viruses infect living cells and make use of the cells energy and material resources to reproduce themselves. In doing so, they take over the cell's metabolism and thereby interfere with the normal processes of the cell, bringing about cell or organism disease or death. The severity of the impact depends upon many factors such as the nature of the virus and the health of the host.

Viruses are very specific to their host cells. For example, the Human Immunodeficiency Virus (HIV) only infects cells of the human immune system, rendering the host organism unable to fight off common infections. Feline leukemia virus (FeLV) has a similar impact, but only on cats, leading to death in about 85% of the cases. The Tobacco Mosaic Virus (TMV) infects and destroys leaves of the tobacco plant, thus causing economic consequences for owners of this crop.

Among the human diseases caused by viruses are the common cold, influenza, measles, genital warts, herpes, and some cancers, such as those cause by the human papillomavirus. Since viruses are not living organisms, they cannot be killed, though they can be destroyed by different things such as exposure to air, chemicals, or immune responses, depending on the virus. Also due to the noncellular nature of virus structure, antibiotics are not effective. However, immunity against viruses can be stimulated through vaccination, and treatment of viral symptoms through medicines is available for most human viruses, including HIV.

Skill 6.3   Relate the structures and functions (e.g. morphology, motility, reproduction and growth, metabolic diversity) of prokaryotic organisms to their behavior and identification

Structure and function dictate behavior and aid in the identification of prokaryotic organisms. Important structural and functional aspects of prokaryotes are morphology, motility, reproduction and growth, and metabolic diversity.

Morphology refers to the shape of a cell. The three main shapes of prokaryotic cells are spheres (cocci), rods (bacilli), and spirals (spirilla). Observation of cell morphology with a microscope can aid in the identification and classification of prokaryotic organisms. The most important aspect of prokaryotic morphology, regardless of the specific shape, is the small size of the cells. Small cells allow for rapid exchange of wastes and nutrients across the cell membrane promoting high metabolic and growth rates.

Motility refers to the ability of an organism to move and its mechanism of movement. While some prokaryotes glide along solid surfaces or use gas vesicles to move in water, the vast majority of prokaryotes move by means of flagella. Motility allows organisms to reach different parts of its environment in the search for favorable conditions. Flagellar structure allows differentiation of Archaea and Bacteria as the two classes of prokaryotes have very different flagella.

In addition, different types of bacteria have flagella positioned in different locations on the cell. The locations of flagella are on the ends (polar), all around (peritrichous), or in a tuft at one end of the cell (lophotrichous).

Most prokaryotes reproduce by binary fission, the growth of a single cell until it divides in two. Because of their small size, most prokaryotes have high growth rates under optimal conditions. Environmental factors greatly influence prokaryotic growth rate. Scientists identify and classify prokaryotes based on their ability or inability to survive and grow in certain conditions. Temperature, pH, water availability, and oxygen levels differentially influence the growth of prokaryotes. For example, certain types of prokaryotes can survive and grow at extremely hot or cold temperatures while most cannot.

Prokaryotes display great metabolic diversity. Autotrophic prokaryotes use carbon dioxide as the sole carbon source in energy metabolism, while heterotrophic prokaryotes require organic carbon sources. More specifically, chemoautotrophs use carbon dioxide as a carbon source and inorganic compounds as an energy source, while chemoheterotrophs use organic compounds as a source of energy and carbon. Photoautotrophs require only light energy and carbon dioxide, while photoheterotrophs require an organic carbon source along with light energy. Examining an unknown organism's metabolism aids in the identification process.

Skill 6.4   Differentiate between the major types of bacterial genetic recombination (e.g. transduction, transformation, and conjugation)

Genetic recombination, the combining of DNA from two individuals into the genome of a single individual, creates genetic diversity within bacterial populations. The three major types of bacterial genetic recombination are transduction, transformation, and conjugation.

Transduction is the transfer of DNA by a phage (a virus that infects bacteria) from one host cell to another. The two types of transduction are generalized and specialized. In *generalized transduction*, a lytic phage hydrolyzes the host DNA and makes new phage capsids containing viral DNA for further distribution. Occasionally, a phage capsid forms around a piece of the host bacterial cell DNA and, when the new phage infects other bacterial cells, it introduces foreign, bacterial DNA to the host genome. In *specialized transduction*, the genome of a lysogenic phage integrates into the host genome as a prophage. Excision of the prophage genome for viral replication will sometimes include parts of the bacterial host DNA near the prophage site. When a virus carrying bacterial DNA infects another host cell, it injects the bacterial DNA along with its own DNA. Specialized transduction transfers only certain genes, those near the site of prophage integration.

Transformation is the alteration of a bacteria's genome by uptake of free, foreign DNA from the surrounding environment. Many bacterial cells have surface proteins that recognize and aid in the uptake of DNA from closely related species. For example, a harmless strain of the bacteria *Streptococcus pneumoniae* can become a pathogenic, pneumonia-causing strain by absorbing genes from the environment released by the pathogenic strain.

Conjugation is the direct transfer of DNA between two temporarily joined bacterial cells. The DNA donor, or "male" cell, attaches to the DNA recipient, or "female" cell, with appendages called pili. The formation of a temporary cytoplasmic bridge allows for the transfer of DNA. The ability to transfer DNA as a donor cell arises from the presence of the F factor, a stretch of DNA incorporated into the bacterial genome or carried as a plasmid. When a cell with the F factor built into its chromosome conjugates, it passes small portions of its chromosomal DNA along with the F factor genes onto its conjugation partner. Such transfers result in recombinant bacteria and an increase in genetic variability.

Skill 6.5   Relate microbial processes and products that are helpful or harmful to human beings and their use in biotechnology

Although bacteria and fungi may cause disease, they are also beneficial for use as medicines and food. Penicillin is derived from a fungus that is capable of destroying the cell wall of bacteria. Most antibiotics work in this way. Some antibiotics can interfere with bacterial DNA replication or can disrupt the bacterial ribosome without affecting the host cells. Viral diseases have been fought through the use of vaccination, where a small amount of the virus is introduced so the immune system is able to recognize it upon later infection. Antibodies are more quickly manufactured when the host has had prior exposure.

The majority of prokaryotes decompose material for use by the environment and other organisms. This can be helpful to humans, as it allows recycling of nutrients into the soil, but it can also be harmful, in that our food can be decomposed by microbes. Salting, pickling, refrigerating, freezing, pasteurizing, and drying are ways we prepare foods to minimize this decomposition by microbes.

Yeast cells ferment and are utilized in baking and wine making. Certain bacterial species are used in producing yogurt and other foods.

Biotechnology includes the use of bacterial cells as factories of gene production, and viruses as carriers of genes between cells.

Competency 7.0   Knowledge of the structural and functional diversity of Protists, Fungi and Plants

Skill 7.1   Identify major types of protists, fungi, and plants

Protists are the earliest eukaryotic descendants of prokaryotes. Protists are found almost anywhere there is water. Protists can be broadly defined as the eukaryotic microorganisms and include the macroscopic algae with only a single tissue type. Most protists have a true (membrane-bound) nucleus, complex organelles (mitochondria, chloroplasts, etc.), aerobic respiration in mitochondria, and undulipodium (cilia) in some life stage. The protists are often grouped as algae (plant)-like, protozoa (animal)-like, or fungus-like, based on the similarity of their lifestyle and characteristics to these more derived groups.

The eukaryotic fungi are the most important decomposers in the biosphere. They break down organic material to be used up by other living organisms. The fungi are characterized by a short lived diploid stage, which cannot be viewed except under a microscope. The structures that are visible to the naked eye are typically puffballs, mushrooms, and shelf fungi that represent the dikaryote form of the fungi. The haploid stages are commonly observed as the absorptive hyphae or as asexual reproductive sporangia.

The **non-vascular plants** represent a grade of evolution characterized by several primitive features for plants: lack of roots, lack of conducting tissues, rely on absorption of water that falls on the plant or they live in a zone of high humidity, and a lack of leaves or have microphylls (in ferns). Groups included are the liverworts, hornworts, and mosses. Each is recognized as a separate division.

The characteristics of **vascular plants** are as follows: synthesis of lignin to give rigidity and strength to cell walls for growing upright, evolution of tracheid cells for water transport and sieve cells for nutrient transport, and the use of underground stems (rhizomes) as a structure from which adventitious roots originate. There are two kinds of vascular plants: non-seeded and seeded. The non-seeded vascular plants divisions include Division Lycophyta – club moses, Division Sphenophyta – horsetails, and Division Pterophyta – ferns. The seeded vascular plants differ from the non-seeded plants by their method of reproduction, which will be discussed later.

The vascular seed plants are divided into two groups, the gymnosperms and the angiosperms. **Gymnosperms** were the first plants to evolve with the use of seeds for reproduction which made them less dependent on water to assist in reproduction. Their seeds and the pollen from the male are carried by the wind. Gymnosperms have cones that protect the seeds.

Gymnosperm divisions include Division Cycadophyta – cycads, Division Ginkgophyta – ginkgo, Division Gnetophyta – gnetophytes, and Division Coniferophyta – conifers.

Angiosperms are the largest group in the plant kingdom. They are the flowering plants and produce true seeds for reproduction. Unlike the seeds of the gymnosperms, the seeds of the angiosperms are surrounded by an ovary. They arose about seventy million years ago when the dinosaurs were disappearing. The land was drying up and the plants' ability to produce seeds that could remain dormant until conditions became acceptable allowed for their success. They also have more advanced vascular tissue and larger leaves for increased photosynthesis. One result of photosynthesis is the release of oxygen, which humans rely on to breathe.

Angiosperms consist of only one division, the Anthrophyta. Angiosperms are divided into monocots and dicots. Monocots have one cotelydon (seed leaf) and parallel veins on their leaves. Their flower petals are in multiples of threes. Dicots have two cotelydons and branching veins on their leaves. Flower petals are in multiples of fours or fives.

Skill 7.2   **Characterize the relationships of protists, fungi, and plants to other living things.**

Protists are entirely unique and diverse group and as such have been difficult for scientists to classify. They're not plants, animals or fungi, but they act enough like them that scientists believe protists paved the way for the evolution of early plants, animals, and fungi. Protists fall into four general subgroups: unicellular algae, protozoa, slime molds, and water molds. Protists may be found in freshwater, saltwater, dirt, and also inside other organisms in a symbiotic relationship. All protists are eukaryotes, and some are photosynthetic while others use phagocytosis (engulfing another cell) to eat. Plant-like algae produce much of the oxygen we breathe; animal-like protozoa (including the famous amoeba) help maintain the balance of microbial life.

Fungi range in size from the single-celled organism we know as yeast to the largest known living organism on Earth — a 3.5-mile-wide mushroom. Fungi are eukaryotic organisms. Many of them may look plant-like, but they are not capable of photosynthesis. Some fungi are useful to humans. Scientists have used several kinds to make antibiotics to fight bacterial infections (ex. Penicillin). These antibiotics are based on natural compounds the fungi produce to compete against bacteria for nutrients and space. We also use *Saccharomyces cerevisiae* to make bread rise and to brew beer. Fungi are decomposers. Unlike plants, fungi do not make their own food energy via photosynthesis, but dine on organic matter like rotting leaves, wood, and other debris. In the process of breaking down organic matter they keep debris from accumulating. In addition, fungi can be helpful to crops. One subgroup, the Mycorrhizae, attach to the roots of plants and help the plants take in nutrients from the soil; nutrients the plants need but couldn't absorb well by themselves. The opposite can also be true: some fungi cause serious plant diseases.

Plants are necessary to other living things because they are at the bottom of the terrestrial food web. They make their own energy utilizing sunlight through the process of photosynthesis. In addition, one of the byproducts of photosynthesis is oxygen, which is of primary importance to breathing mammals.

Skill 7.3   Distinguish between the structures and functions of various plant tissues.

The three basic classifications of plant tissue are ground, dermal, and vascular.

Ground tissue occupies the space between the dermal and vascular tissue and synthesizes organic compounds, supports the plant, and provides storage for the plant. Ground tissue consists of parenchyma, sclerenchyma, and collenchyma tissue. Parenchyma is the main component of ground tissue and is the progenitor of all other tissue. Parenchyma cells are living at maturity, non-specialized, and perform most of the plant metabolic activities. Sclerenchyma tissue consists of non-living cells that are very rigid and have secondary walls and a hardening agent that provide support to the plant. Finally, collenchyma tissue consists of living cells that are somewhat rigid and provide support to growing plants, but do not possess secondary walls or a hardening agent.

Dermal tissue is the outermost layer of plant leaves, stems, fruits, seeds, and roots. Dermal tissue interacts with the environment and its functions include gas exchange, light passage, and pathogen recognition. One subtype of dermal tissue is the epidermis. The epidermis is usually a single layer of unspecialized cells, either parenchyma or sclerenchyma. In plants that undergo secondary growth, another subtype of dermal tissue called periderm replaces the epidermis on the stems and roots. Periderm, also called bark, consists of parenchyma, sclerenchyma, and cork cells. The function of the periderm is to prevent excess water loss, protect against pathogens, and provide insulation to the plant.

Vascular tissue consists of xylem and phloem and facilitates the transport of water and nutrients throughout the plant. Xylem is a sclerenchyma tissue that conducts water and transports minerals from the soil up the plant. The two types of conductive cells in the xylem are tracheids and vessels. Vessels are much larger in diameter and serve as the major pipes in the water transport system of the plant. Phloem is a parenchyma tissue that functions in the transport of sugars, amino acids, and other small molecules from the leaves to the rest of the plant. The two types of specialized cells in the phloem are sieve elements and companion cells. Sieve elements possess thin cell walls and pores that facilitate transfer of nutrients. Companion cells accompany sieve element cells and help maintain the life functions of the sieve cells.

Skill 7.4 Identify the characteristics of vascular and nonvascular plants and relate these characteristics to adaptations allowing these plants to broaden their ecological niches

The **non-vascular plants** represent a grade of evolution characterized by several primitive features for plants: lack of roots, lack of conducting tissues, rely on absorption of water that falls on the plant or they live in a zone of high humidity, and a lack of leaves or have microphylls (in ferns). Groups included are the liverworts, hornworts, and mosses. Each is recognized as a separate division.

The characteristics of **vascular plants** are as follows: synthesis of lignin to give rigidity and strength to cell walls for growing upright, evolution of tracheid cells for water transport (xylem tissue) and sieve cells for nutrient transport (phloem tissue), and the use of underground stems (rhizomes) as a structure from which adventitious roots originate. There are two kinds of vascular plants: non-seeded and seeded. The non-seeded vascular plants divisions include Division Lycophyta – club mosses, Division Sphenophyta – horsetails, and Division Pterophyta – ferns. The seeded vascular plants differ from the non-seeded plants by their method of reproduction, which will be discussed later.

Skill 7.5 Identify the functions and survival advantages of the major organs of angiosperms and gymnosperms

The vascular seed plants are divided into two groups, the gymnosperms and the angiosperms. **Gymnosperms** were the first plants to evolve with the use of seeds for reproduction which made them less dependent on water to assist in reproduction. Their seeds and the pollen from the male are carried by the wind. Gymnosperms have cones that protect the seeds. Gymnosperm divisions include Division Cycadophyta – cycads, Division Ginkgophyta – ginkgo, Division Gnetophyta – gnetophytes, and Division Coniferophyta – conifers.

Angiosperms are the largest group in the plant kingdom. They are the flowering plants and produce true seeds for reproduction. They arose about seventy million years ago when the dinosaurs were disappearing. The land was drying up and the plants' ability to produce seeds that could remain dormant until conditions became acceptable allowed for their success. They also have more advanced vascular tissue and larger leaves for increased photosynthesis.

Skill 7.6   Distinguish between the structures of monocots and dicots (e.g., seeds, vascular bundles, venation, flower parts)

Monocots have one cotelydon (seed leaf) and parallel veins on their leaves. Their flower petals are in multiples of threes. Vascular bundles are scattered across the stem's diameter. Dicots have two cotelydons and branching veins on their leaves. Flower petals are in multiples of fours or fives. Vascular bundles are arranged in a ring near the outside of the stem's diameter.

Skill 7.7   Identify the major mechanisms (e.g., transport, storage, conservation) in plants and evaluate the survival advantages these mechanisms give to different groups of plants

Plants require adaptations that allow them to absorb light for photosynthesis. Since they are unable to move about, they must evolve methods to allow them to successfully reproduce. As time passed, the plants moved from a water environment to the land. Advantages of life on land included more available light and a higher concentration of carbon dioxide. Originally, there were no herbivores and less competition for space on land. Plants had to evolve methods of support, reproduction, transport of internal fluids, respiration, and conservation of water once they moved to land. Reproduction by plants is accomplished through alternation of generations. Simply stated, a haploid stage in the plant's life history alternates with a diploid stage. A general evolutionary trend in plants favors methods such as pollen and seeds and the increasing importance of the sporophyte generation, which all assist with reproduction on land. A division of labor among plant tissues (phloem and xylem) evolved in order to get food from the leaves to all parts of the plant, and water and minerals from the earth as described above.. A wax cuticle is produced to prevent the loss of water. Leaves enabled plants to capture light and carbon dioxide for photosynthesis. Stomata provide openings on the underside of leaves for oxygen to move in or out of the plant and for carbon dioxide to move in. A method of anchorage (roots) evolved. The polymer lignin evolved to give tremendous strength to plants.

Skill 7.8   Analyze the role of major plant growth regulators

Plant growth regulators, or plant hormones, are chemicals secreted internally by plants that regulate growth and development. Plant growth regulators are present in low concentrations, produced in specific locations, and often act on cells at other locations. The five major classes of plant growth regulators are: (1) auxins, (2) abscisic acid, (3) gibberellins, (4) ethylene, and (5) cytokinins.

Auxins play a major role in many growth and behavioral processes. Auxins promote cell elongation in growing shoot tips and cell expansion in swelling roots and fruits. In addition, auxins promote apical dominance, the tendency of the main stem of plants to grow more strongly than the side stems. Other effects of auxins include phototropism (plants bending toward light), stimulation of ethylene synthesis, stimulation of cell division, and inhibition of abscission (i.e. leaf shedding).

Abscisic acid (ABA) plays an important role in promoting dormancy, inhibiting growth, and responding to water stress. High levels of ABA induce seed dormancy and inhibit germination. In addition, ABA builds up during water stress promoting closing of stomata (i.e. pores). Finally, ABA prevents fruit ripening.

Gibberellins have a wide range of effects. The most important effect of gibberellins is stem elongation. Gibberellins also promote flower and fruit formation.   Finally, gibberellins stimulate the growth and development of seeds.

Ethylene is the major growth factor involved in fruit ripening and plant part abscission. Ethylene also induces seed germination, root hair growth, and flowering.

Cytokinins play a key role in cell division. Cytokinins generally promote shoot development and inhibit root development. In addition, cytokinins delay senescence in flowers and fruits, promote chlorophyll production, and promote photosynthesis.

 Skill 7.9   Apply concepts of major methods of reproduction in plants, including dispersal mechanisms

Reproduction by plants is accomplished through alternation of generations. Simply stated, a haploid stage in the plants life history alternates with a diploid stage. The cells of the diploid sporophyte divide by meiosis to reduce the chromosome number to the haploid number in specialized cells called spores. These give rise to the gametophyte generation. The haploid gametophytes undergo mitosis to produce gametes (sperm and eggs). Then, the haploid gametes fertilize to return to the diploid sporophyte stage.

The non-vascular plants and the vascular non-seeded plants need environmental water to reproduce, since the male gamete is a flagellated swimming sperm. . In most of these plants, the gametophyte stage is more predominant. In the seed plants, the Gymnosperms and the Angiosperms, the gametophyte is a small attached part of the sporophyte, The male gamete is pollen, which unlike sperm, can survive dry conditions and be airborne, so environmental water is not required for fertilization, though insect and other pollinators sometimes are so required in some Angiosperms.

Angiosperms are the most numerous and are therefore the main focus of reproduction in this section. The sporophyte is the dominant phase in reproduction. Angiosperm reproductive structures are the flowers.

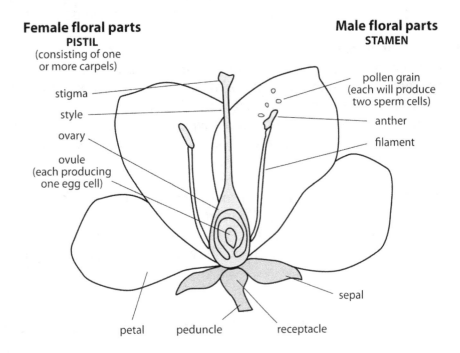

The male gametophytes are pollen grains and the female gametophytes are embryo sacs that are inside of the ovules. The male pollen grains are formed in the anthers at the tips of the stamens. The female ovules are enclosed by the ovaries. Therefore, the stamen is the reproductive organ of the male and the carpel is the reproductive organ of the female.

In a process called **pollination**, the pollen grains are released from the anthers and carried by animals and the wind and land on the carpels. The sperm is released to fertilize the eggs. Angiosperms reproduce through a method of double fertilization. An ovum is fertilized by two sperm. One sperm produces the new plant and the other forms the food supply for the developing plant (endosperm). The ovule develops into a seed and the ovary develops into a fruit. The fruit is then carried by wind or animals and the seeds are dispersed to form new plants.

The development of the egg to form a plant occurs in three stages: growth; morphogenesis; the development of form; and cellular differentiation, the acquisition of a cell's specific structure and function.

Skill 7.10 Analyze patterns of alternation of generations in various groups of plants and algae

Alternation of generations refers to the life cycle of some plants and algae consisting of haploid and diploid phases. In the diploid (or sporophyte) phase, each cell of the organism has two complete sets of chromosomes. In the haploid, or gametophyte, phase, each cell of the organism has only one set of chromosomes. The haploid plant produces gametes by mitosis that combine to form the diploid sporophyte. The diploid sporophyte produces spores by meiosis that develop into haploid gametophytes.

In primitive plants like bryophytes (e.g. mosses), both the gametophyte and sporophyte plants are conspicuous and the gametophyte phase is actually the dominant plant observed. This contrasts greatly with less primitive plants, gymnosperms and angiosperms, in which the sporophyte plant is dominant. The gametophyte stage of gymnosperms and angiosperms consists of a few cells nourished by the sporophyte. In other words, the gametophyte is not a free-living organism and the plants observed in nature are the diploid generation.

Many types of algae (e.g. red, green, and brown algae) exhibit isomorphic alternation of generations. In isomorphic alternation of generations, the gametophyte and sporophyte generations are identical. Conversely, some algae and all plants exhibit heteromorphic isolation of generations in which the sporophyte and gametophyte generations are structurally different.

Competency 8.0 Knowledge of the structural and functional diversity of animals

Skill 8.1 Relate the structures of major animal tissue types to their function

Animal tissue becomes specialized during development. The ectoderm (outer layer) becomes the epidermis or skin. The mesoderm (middle layer) becomes muscles and other organs beside the gut. The endoderm (inner layer) becomes the gut, also called the archenteron.

The developed tissue types each have specialized structures that directly support their specific function. For example, muscle tissue is composed of strands of actin and myosin protein that give muscle its unique ability to contract. Nervous tissue however is completely different, having extremely long cell parts covered with a myelin sheath to allow for the transmission of nervous impulses.

Skill 8.2 Identify major animal body plans (e.g., symmetry, coelomic character, embryonic origin)

Sponges are the simplest animals and lack true tissue. They exhibit no symmetry.

Diploblastic animals have only two germ layers: the ectoderm and endoderm. They have no true digestive system. Diploblastic animals include the Cnideria (jellyfish). They exhibit radial symmetry.

Triploblastic animals have all three germ layers. Triploblastic animals can be further divided into:

Acoelomates - have no defined body cavity. An example is the flatworm (Platyhelminthe), which must absorb food from a host's digestive system.

Pseudocoelomates - have a body cavity but it is not lined by tissue from the mesoderm. An example is the roundworm (Nematoda).

Coelomates - have a true fluid filled body cavity called a coelom derived from the mesoderm. Coelomates can further be divided into **protostomes** and **deuterostomes**. In the development of protostomes, the first opening becomes the mouth and the second opening becomes the anus. The mesoderm splits to form the coelom. In the development of deuterostomes, the mouth develops from the second opening and the anus from the first opening. The mesoderm hollows out to become the coelom. Protostomes include animals in phyla Mollusca, Annelida and Arthropoda. Deuterostomes include animals in phyla Ehinodermata and Vertebrata.

Development is defined as a change in form. Animals go through several stages of development after fertilization of the egg cell:

Cleavage - the first divisions of the fertilized egg. Cleavage continues until the egg becomes a blastula.

Blastula - the blastula is a hollow ball of undifferentiated cells.
Gastrulation - this is the time of tissue differentiation into the separate germ layers, the endoderm, mesoderm and ectoderm.

Neuralation - development of the nervous system.

Organogenesis - the development of the various organs of the body.

Skill 8.3 **Relate the processes of animal growth and development to early embryological development (e.g. embryonic induction, ontogeny recapitulating phylogeny)**

Embryonic induction is the process in which developing embryonic tissue influences surrounding tissue, thus changing the responding tissue's pattern of differentiation. For example, the development of neural tissue in a developing embryo relies on embryonic induction. Some ectodermal cells develop into neuroectoderm, the precursor of the nervous system, when signaled by tissue in the notochord, the central axis of the embryo. Embryonic induction is comparable to hormonal activity in animal growth and development. Hormones released by cells often stimulate growth and development of other cells.

Studying ontogeny, the origin and development of an organism from fertilization to adulthood, provides insight into the evolutionary history of an organism. In fact, some scientists in the 19th century believed ontogeny recapitulated phylogeny. In other words, an organism's evolutionary development was preserved, step-by-step in the development of the embryo. While this theory has been discredited, some ancestral characteristics do appear in the developmental stages of embryos. For example, both chicken and human embryos pass through a stage where they have slits that are identical to the gills of fish. In addition, human embryos possess tail-like structures at one point in their development. The appearance of such traits helps link species to their evolutionary ancestors.

Skill 8.4 **Relate the structures to functions of circulatory and respiratory systems**

The function of the closed circulatory system (**cardiovascular system**) is to carry oxygenated blood and nutrients to all cells of the body and return carbon dioxide waste to be expelled from the lungs. The heart, blood vessels, and blood make up the cardiovascular system. The structure of the heart is shown below:

The atria are the chambers that receive blood returning to the heart and the ventricles are the chambers that pump blood out of the heart. There are four valves, two atrioventricular (AV) valves and two semilunar valves. The AV valves are located between each atrium and ventricle. The contraction of the ventricles closes the AV valve to keep blood from flowing back into the atria. The semilunar valves are located where the aorta leaves the left ventricle and the pulmonary artery leaves the right ventricle. The semilunar valves are opened by ventricular contraction to allow blood to be pumped out into the arteries and closed by the relaxation of the ventricles.

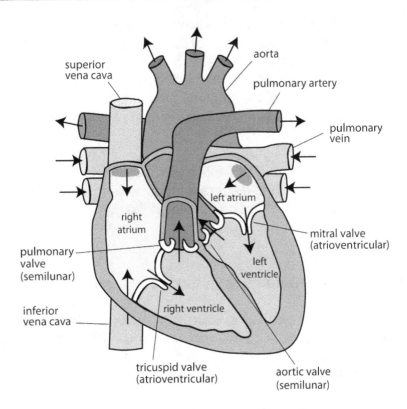

The cardiac output is the volume of blood per minute that the left ventricle pumps. This output depends on the heart rate and stroke volume. The **heart rate** is the number of times the heart beats per minute and the **stroke volume** is the amount of blood pumped by the left ventricle each time it contracts. Humans have an average cardiac output of about 5.25 L/min. Heavy exercise can increase cardiac output up to five times. Epinephrine and increased body temperature also increase heart rate and thus the cardiac output. Cardiac muscle can contract without any signal from the nervous system. It is the sinoatrial node that is the pacemaker of the heart. It is located on the wall of the right atrium and generates electrical impulses that make the cardiac muscle cells contract in unison. The atrioventricular node shortly delays the electrical impulse to ensure the atria empty before the ventricles contract.

The lungs are the respiratory surface of the human respiratory system. A dense net of capillaries contained just beneath the epithelium form the respiratory surface. The surface area of the epithelium is about 100m² in humans. Based on the surface area, the volume of air inhaled and exhaled is the tidal volume.

This is normally about 500mL in adults. Vital capacity is the maximum volume the lungs can inhale and exhale. This is usually around 3400mL.

The respiratory system functions in the gas exchange of oxygen and carbon dioxide waste. It delivers oxygen to the bloodstream and picks up carbon dioxide for release out of the body. Air enters the mouth and nose, where it is warmed, moistened and filtered of dust and particles. Cilia in the trachea trap unwanted material in mucus, which can be expelled. The trachea splits into two bronchial tubes and the bronchial tubes divide into smaller and smaller bronchioles in the lungs. The internal surface of the lung is composed of alveoli, which are thin walled air sacs. These allow for a large surface area for gas exchange. The alveoli are lined with capillaries. Oxygen diffuses into the bloodstream and carbon dioxide diffuses out of the capillaries to be exhaled out of the lungs due to partial pressure. The oxygenated blood is carried to the heart and delivered to all parts of the body by hemoglobin, a protein consisting of iron.

The thoracic cavity holds the lungs. The diaphragm muscle below the lungs is an adaptation that makes inhalation possible. As the diaphragm muscle flattens out, the volume of the thoracic cavity increases and inhalation occurs. When the diaphragm relaxes, exhalation occurs.

Skill 8.5 **Relate the structures to functions of excretory and digestive systems**

The kidneys are the primary organ in the excretory system. The pair of kidneys in humans are about 10cm long each. They receive about 20% of the blood pumped with each heartbeat despite their small size. The function of the excretory system is to rid the blood of nitrogenous wastes in the form of urea.

The functional unit of excretion is the nephron, which makes up the kidneys. The Bowman's capsule contains the glomerulus, a tightly packed group of capillaries in the nephron. The glomerulus is under high pressure. Water, urea, salts, and other fluids leak out due to pressure into the Bowman's capsule. This fluid waste (filtrate) passes through the three regions of the nephron: the proximal convoluted tubule, the loop of Henle, and the distal tubule. In the proximal convoluted tubule, unwanted molecules are secreted into the filtrate. In the loop of Henle, salt is actively pumped out of the tube and much water is lost due to the hyperosmotic inner part (medulla) of the kidney. As the fluid enters the distal tubule, more water is reabsorbed. Urine forms in the collecting duct that leads to the ureter then to the bladder where it is stored. Urine is passed from the bladder through the urethra. The amount of water reabsorbed back into the body is dependent upon how much water or fluids an individual has consumed. Urine can be very dilute or very concentrated if dehydration is present.

The structures of the kidney and nephron are as follows:

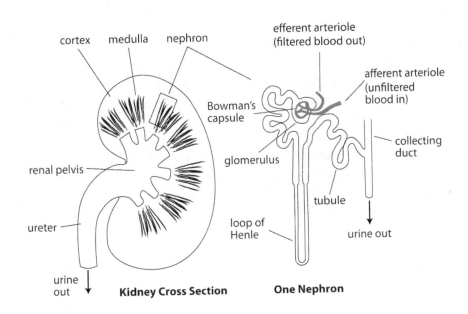

Kidney Cross Section   **One Nephron**

The function of the digestive system is to break food down into nutrients and absorb it into the blood stream where it can be delivered to all cells of the body for use in cellular respiration.

Essential nutrients are those nutrients that the body needs but cannot make. There are four groups of essential nutrients: essential amino acids, essential fatty acids, vitamins, and minerals.

There are about eight essential amino acids humans need. A lack of these amino acids results in protein deficiency. There are only a few essential fatty acids.

Vitamins are organic molecules essential for a nutritionally adequate diet. Thirteen vitamins essential to humans have been identified. There are two groups of vitamins: water soluble (includes the vitamin B complex and vitamin C) and water insoluble (vitamins A, D and K). Vitamin deficiencies can cause severe problems.

Unlike vitamins, minerals are inorganic molecules. Calcium is needed for bone construction and maintenance. Iron is important in cellular respiration and is a big component of hemoglobin.

Carbohydrates, fats, and proteins are fuel for the generation of ATP. Water is necessary to keep the body hydrated. The importance of water was discussed in previous sections.

The teeth and saliva begin digestion by breaking food down into smaller pieces and lubricating it so it can be swallowed. Some carbohydrate digestion begins in the mouth due to the enzyme amylase. The lips, cheeks, and tongue form a bolus or ball of food. It is carried down the pharynx by the process of peristalsis (wave-like contractions) and enters the stomach through the sphincter, which closes to keep food from going back up. In the stomach, pepsinogen and hydrochloric acid form pepsin, the enzyme that hydrolyzes proteins. The food is broken down further by this chemical action and is churned into acid chyme. The pyloric sphincter muscle opens to allow the food to enter the small intestine. Most nutrient absorption occurs in the small intestine. Its large surface area, accomplished by its length and protrusions called villi and microvilli, allow for a great absorptive surface into the bloodstream. Chyme is neutralized after coming from the acidic stomach to allow the enzymes found there to function. Accessory organs function in the production of necessary enzymes and bile. The pancreas makes many enzymes to break down food in the small intestine. The liver makes bile, which breaks down and emulsifies fatty acids. Any food left after the trip through the small intestine enters the large intestine. The large intestine functions to reabsorb water and produce vitamin K. The feces, or remaining waste, are passed out through the anus.

Skill 8.6 Relate the structures to functions of endocrine and nervous systems

The function of the **endocrine system** is to manufacture proteins called hormones. **Hormones** are released into the bloodstream and are carried to a target tissue where they stimulate an action. There are two classes of hormones. Steroid hormones come from cholesterol and include the sex hormones. Peptide hormones are derived from amino acids. Hormones are specific and fit receptors on the target tissue cell surface. The receptor activates an enzyme that converts ATP to cyclic AMP. Cyclic AMP (cAMP) is a second messenger from the cell membrane to the nucleus. The genes found in the nucleus turn on or off to cause a specific response.

Hormones are secreted by endocrine cells which make up endocrine glands. The major endocrine glands and their hormones are as follows:

Hypothalamus – located in the lower brain; signals the pituitary gland.

Pituitary gland – located at the base of the hypothalamus; releases growth hormones and antidiuretic hormone (retention of water in kidneys).

Thyroid gland – located on the trachea; lowers blood calcium levels (calcitonin) and maintains metabolic processes (thyroxin).

Gonads – the testes of the male and the ovaries of the female; testes release androgens to support sperm formation and ovaries release estrogens to stimulate uterine lining growth and progesterone to promote uterine lining growth.

Pancreas – secretes insulin to lower blood glucose levels and glucagon to raise blood glucose levels.

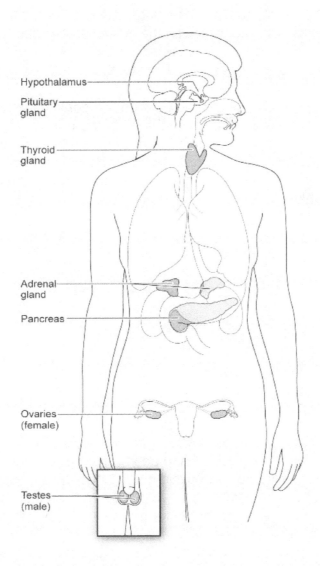

The **central nervous system (CNS)** consists of the brain and spinal cord. The CNS is responsible for the body's response to environmental stimulation. The spinal cord is located inside the spine. It sends out motor commands for movement in response to stimuli. The brain is where responses to more complex stimuli occur. The meninges are the connective tissues that protect the CNS. The CNS contains fluid filled spaces called ventricles. These ventricles are filled when cerebrospinal fluid which is formed in the brain. This fluid cushions the brain and circulates nutrients, white blood cells, and hormones. The CNS's response to stimuli may be a reflex. The reflex is an unconscious, automatic response of the spinal cord. In contrast, a response to a stimulus might be mediated by multiple higher brain processes of sensing and conscious thought.

The **peripheral nervous system (PNS)** consists of the nerves that connect the CNS to the rest of the body. The sensory division brings information to the CNS from sensory receptors and the motor division sends signals from the CNS to effector cells. The motor division consists of somatic nervous system and the autonomic nervous system. The somatic nervous system is controlled consciously in response to external stimuli. The autonomic nervous system is unconsciously controlled by the hypothalamus of the brain to regulate the internal environment. This system is responsible for the movement of smooth and cardiac muscles as well as the muscles for other organ systems.

The **neuron** is the basic unit of the nervous system. It consists of an axon, which carries impulses away from the cell body to the tip of the neuron; the dendrite, which carries impulses toward the cell body; and the cell body, which contains the nucleus. Synapses are spaces between neurons. Chemicals called neurotransmitters are found close to the synapse. The myelin sheath is composed of Schwann cells and covers the neurons, providing insulation.

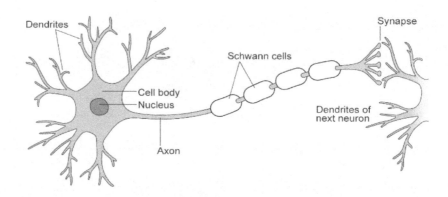

Nerve action depends on depolarization and an imbalance of electrical charges across the neuron. A polarized nerve has a positive charge outside the neuron. A depolarized nerve has a negative charge outside the neuron. Neurotransmitters turn off the sodium pump which results in depolarization of the membrane. This wave of depolarization (as it moves from neuron to neuron) carries an electrical impulse. This is actually a wave of opening and closing gates that allows for the flow of ions across the synapse. Nerves have an action potential. There is a threshold of the level of chemicals that must be met or exceeded in order for muscles to respond. This is called the "all or none" response.

 Skill 8.7 **Relate the structures to functions of integumentary and musculoskeletal systems**

The skin consists of two distinct layers. The epidermis is the thinner outer layer and the dermis is the thicker inner layer. Layers of tightly packed epithelial cells make up the epidermis. The tight packaging of the epithelial cells supports the skin's function as a protective barrier against infection.

The top layer of the epidermis consists of dead skin cells and is filled with keratin, a waterproofing protein. The dermis layer consists of connective tissue. It contains blood vessels, hair follicles, sweat glands, and sebaceous glands. An oily secretion called sebum, produced by the sebaceous gland, is released to the outer epidermis through the hair follicles. Sebum maintains the pH of the skin between 3 and 5, which inhibits most microorganism growth.

The muscular system's function is for movement. There are three types of muscle tissue. **Skeletal muscle** is voluntary. These muscles are attached to bones and are responsible for their movement. Skeletal muscle consists of long fibers and is striated due to the repeating patterns of the myofilaments (made of the proteins actin and myosin) that make up the fibers.

Cardiac muscle is found in the heart. Cardiac muscle is striated like skeletal muscle, but differs in that plasma membrane of the cardiac muscle causes the muscle to beat even when away from the heart. The action potentials of cardiac and skeletal muscles also differ.

Smooth muscle is involuntary. It is found in organs and enable functions such as digestion and respiration. Unlike skeletal and cardiac muscle, smooth muscle is not striated. Smooth muscle has less myosin and does not generate as much tension as the striated muscles.

A nerve impulse strikes a muscle fiber. This causes calcium ions to flood the sarcomere. Calcium ions allow ATP to expend energy. The myosin fibers creep along the actin, causing the muscle to contract. Once the nerve impulse has passed, calcium is pumped out and the contraction ends.

The axial skeleton consists of the bones of the skull and vertebrae. The appendicular skeleton consists of the bones of the legs, arms and tail, and shoulder and pelvic girdles. Bone is a connective tissue. Parts of the bone include compact bone which gives strength, spongy bone which contains red marrow to make blood cells, yellow marrow in the center of long bones to store fat cells, and the periosteum which is the protective covering on the outside of the bone.

A joint is defined as a place where two bones meet. Joints enable movement. Ligaments attach bone to bone. Tendons attach bone to muscle. Joints allow great flexibility in movement.

There are three types of joints:

1. Ball and socket – allows for rotation movement. An example is the joint between the shoulder and the humerus. This joint allows humans to move their arms and legs in many different ways.

2. Hinge – movement is restricted to a single plane. An example is the joint between the humerus and the ulna.

3. Pivot – allows for the rotation of the forearm at the elbow and the hands at the wrist.

Skill 8.8   Relate the structures to functions of reproductive systems

Hormones regulate sexual maturation in humans. Humans cannot reproduce until about the puberty age of 8-14, depending on the individual. The hypothalamus begins secreting hormones that help mature the reproductive system and cause development of the secondary sex characteristics. Reproductive maturity in girls occurs with the first ovulation (usually but not necessarily near the time of the first menstruation) and occurs in boys with the first ejaculation of viable sperm.

Hormones also regulate reproduction. In males, the primary sex hormones are the androgens, testosterone being the most important. The androgens are produced in the testes and are responsible for the primary and secondary sex characteristics of the male. Female hormone patterns are cyclic and complex. Most women have a reproductive cycle length of about 28 days. The menstrual cycle is specific to the changes in the uterus. The ovarian cycle results in ovulation and occurs in parallel with the menstrual cycle. This parallelism is regulated by hormones. Five hormones participate in this regulation, most notably estrogen and progesterone. Estrogen and progesterone play an important role in the signaling to the uterus and the development and maintenance of the endometruim. Estrogens are also responsible for the secondary sex characteristics of females.

Gametogenesis is the production of the sperm and egg cells. **Spermatogenesis** begins at puberty in the male. One spermatogonia, the diploid precursor of sperm, produces four sperm. The sperm mature in the seminiferous tubules located in the testes. **Oogenesis**, the production of egg cells (ova), is usually complete by the birth of a female. Egg cells are not released until ovulation begins at puberty. Meiosis forms one ovum with all the cytoplasm and three polar bodies that are reabsorbed by the body. The ova are stored in the ovaries and released each month from puberty to menopause.

Sperm are stored in the seminiferous tubules in the testes where they mature. Mature sperm are found in the epididymis located on top of the testes. During ejaculation, the sperm travel up the **vas deferens** where they mix with semen made in the prostate and seminal vesicles and travel out the urethra.

Ovulation releases the egg into the fallopian tubes that are ciliated to move the egg along. Fertilization of the egg by the sperm normally occurs in the fallopian tube.

If pregnancy does not occur, the egg passes through the uterus and is expelled through the vagina during menstruation. Levels of progesterone and estrogen stimulate menstruation and are affected by the implantation of a fertilized egg so menstruation will not occur.

Skill 8.9   Relate the structures to functions of the human immune system

The immune system is responsible for defending the body against foreign invaders. There are two defense mechanisms: non- specific and specific.

The **non-specific** immune mechanism has two lines of defenses. The first line of defense is the physical barriers of the body. These include the skin and mucous membranes. The skin prevents the penetration of bacteria and viruses as long as there are no abrasions on the skin. Mucous membranes form a protective barrier around the digestive, respiratory, and genitourinary tracts. Also, the pH of the skin and mucous membranes inhibit the growth of many microbes. Mucous secretions (tears and saliva) wash away many microbes and contain lysozyme that kills microbes.

The second line of defense includes white blood cells and the inflammatory response. **Phagocytosis** is the ingestion of foreign particles. Neutrophils make up about seventy percent of all white blood cells. Monocytes mature to become macrophages which are the largest phagocytic cells. Eosinophils are also phagocytic. Natural killer cells destroy the body's own infected cells instead of the invading the microbe directly.

The other second line of defense is the inflammatory response. The blood supply to the injured area is increased, causing redness and heat. Swelling also typically occurs with inflammation. Histamine is released by basophils and mast cells when the cells are injured. This triggers the inflammatory response.

The **specific** immune mechanism recognizes specific foreign material and responds by destroying the invader. These mechanisms are specific and diverse. They are able to recognize individual pathogens. An **antigen** is any foreign particle that elicits an immune response. An **antibody** is manufactured by the body and recognizes and latches onto antigens, hopefully destroying them. They also have recognition of foreign material versus the self. Memory of the invaders provides immunity upon further exposure.

See **Skill 3.10** for more information.

Skill 8.10   Analyze the interconnectedness of animal organ systems

Groups of related organs are organ systems. Organ systems consist of organs working together to perform a common function. The commonly recognized organ systems of animals include the reproductive system, nervous system, circulatory system, respiratory system, lymphatic system (immune system), endocrine system, urinary system, muscular system, digestive system, integumentary system, and skeletal system. In addition, organ systems are interconnected and a single system rarely works alone to complete a task.

One obvious example of the interconnectedness of organ systems is the relationship between the circulatory and respiratory systems. As blood circulates through the organs of the circulatory systems, it is re-oxygenated in the lungs of the respiratory system. Another example is the influence of the endocrine system on other organ systems. Hormones released by the endocrine system greatly influence processes of many organ systems including the nervous and reproductive systems.

In addition, bodily response to infection is a coordinated effort of the lymphatic (immune system) and circulatory systems. The lymphatic system produces specialized immune cells, filters out disease-causing organisms, and removes fluid waste from in and around tissue. The lymphatic system utilizes capillary structures of the circulatory system and interacts with blood cells in a coordinated response to infection. Finally, the muscular and skeletal systems are closely related. Skeletal muscles attach to the bones of the skeleton and drive movement of the body.

Skill 8.11 Analyze the effects of feedback loops in human systems (e.g. classical vertebrate hormones, fight or flight)

Feedback loops in human systems serve to regulate bodily functions in relation to environmental conditions. Positive feedback loops enhance the body's response to external stimuli and promote processes that involve rapid deviation from the initial state. Positive feedback loops function in stress response and the regulation of growth and development. For example, lactation involves a positive feedback loop in that the stimulus (infant sucking) results in increased milk flow and production.

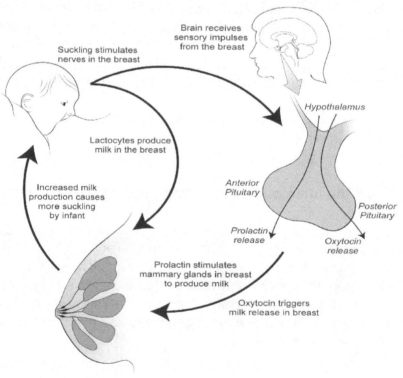

Negative feedback loops help maintain stability in spite of external or internal environmental changes and function in homeostasis. For example, negative feedback loops function in the regulation of blood glucose levels and the maintenance of body temperature.

Feedback loops regulate the secretion of classical vertebrate hormones in humans. The pituitary gland and hypothalamus respond to varying levels of hormones by increasing or decreasing production and secretion. High levels of a hormone cause down-regulation of the production and secretion pathways, while low levels of a hormone cause up-regulation of the production and secretion pathways.

"Fight or flight" refers to the human body's response to stress or danger. Briefly, as a response to an environmental stressor, the hypothalamus releases a hormone that acts on the pituitary gland, triggering the release of another hormone, adrenocorticotropin (ACTH), into the bloodstream. ACTH then signals the adrenal glands to release the hormones cortisol, epinephrine, and norepinephrine. These three hormones act to ready the body to respond to a threat by increasing blood pressure and heart rate, speeding reaction time, diverting blood to the muscles, and releasing glucose for use by the muscles and brain. The stress-response hormones also down-regulate growth, development, and other non-essential functions. Finally, cortisol completes the "fight or flight" feedback loop by acting on the hypothalamus to stop hormonal production after the threat has passed.

Skill 8.12   Identify aspects of animal social behavior (e.g., communication and signals, dominance hierarchy, territoriality, aggression, courtship, innate and learned behavior)

Animal behavior is responsible for courtship leading to mating, communication between individuals and groups of the same or different species, territoriality, and aggression between animals and dominance within a group. Behaviors may include body posture, mating calls, display of feathers or fur, coloration or bearing of teeth and claws.

Innate behaviors are inborn or instinctual. An environmental stimulus such as the length of day or temperature results in a behavior. Hibernation among some animals is an innate behavior. **Learned behavior** is modified due to past experience. Some behaviors have both innate and learned elements.

Competency 9.0   Knowledge of ecological principles and processes

Skill 9.1   Distinguish between individuals, populations, communities, ecosystems, biomes, and the biosphere

A **population** is a group of individuals of one species that live in the same general area. A **community** is a collection of populations that live together in a given area; these populations would all be a part of a food web. An **ecosystem** is a community of living things, plus all the nonliving (abiotic) factors in the environment such as water, soil, temperature, acidity, rainfall, and shape of the land. **Biomes** are ecosystems that are typical of broad geographic regions, and can be characterized as having certain weather patterns, temperatures, and plant and animal species. Specific biomes include: freshwater aquatic, marine aquatic, desert, tundra, deciduous forest, grassland, and rain forest. The **biosphere** is the sum total of all biomes on Earth; it is the planet's thin layer of living things and the nonliving things that support life on and in land and water.

Skill 9.2   Analyze the relationship between organisms and their niches

Within an ecosystem are many species living among other species and abiotic factors in the environment. Each organism type is said to fill a niche within the ecosystem. The niche is the sum total of interactions that organism has with its environment: the food it uses, the resources it exploits for shelter and reproduction, the competitive pressures it exerts on other species, etc. Each niche is understood to be unique to that species within that ecosystem; in other words, two species would not occupy the same niche, since one would outcompete the other. Therefore, the number of different species is equal to the number of niches in an ecosystem.

Skill 9.3   Analyze the roles of organisms in the major biogeochemical cycles and processes

Biogeochemical cycles are nutrient cycles that involve both biotic and abiotic factors.

Water cycle - 2% of all the available water is fixed and unavailable in ice or the bodies of organisms. Available water includes surface water (lakes, ocean, rivers) and ground water (aquifers, wells) 96% of all available water is from ground water. The water cycle is driven by heat energy, which came from solar energy. Water is recycled through the processes of evaporation and precipitation. The water present now is the water that has been here since our atmosphere formed. Since living things contain a lot of water, water is also cycling through all living things. For example, a tree absorbs water from the soil, passes it through its vascular tissue, and then releases it into the air through its stomata in the process of transpiration.

Carbon cycle - Ten percent of all available carbon in the air (from carbon dioxide gas) is fixed by photosynthesis. Plants fix carbon in the form of glucose, animals eat the plants and are able to obtain their source of carbon. When living cells release carbon dioxide through respiration, the carbon is cycled back into the atmosphere as carbon dioxide. Since plant cells use both photosynthesis and respiration, the carbon cycle is completed and plants could survive without animals. However, the reverse is not true since animals cannot photosynthesize to create organic carbon molecules (food).

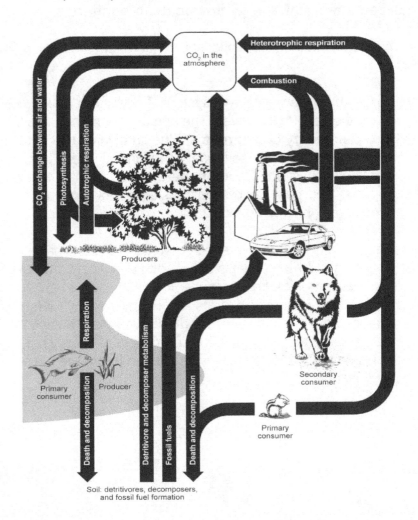

Nitrogen cycle - Eighty percent of the atmosphere is in the form of nitrogen gas. Nitrogen must be fixed and taken out of the gaseous form to be incorporated into an organism. Only a few genera of bacteria have the correct enzymes to break the triple bond between nitrogen atoms in a process called nitrogen fixation. These bacteria live within the roots of legumes (peas, beans, alfalfa) and add nitrogen to the soil in the form of ammonia, nitrates, or nitrites that can be taken up by the plant. Nitrogen is necessary to make amino acids and the nitrogenous bases of DNA. When organisms decompose, their nitrogenous compounds are broken down in many steps until finally denitrifying bacteria break down nitrates to release nitrogen gas back into the atmosphere, thus completing the cycle.

Phosphorus cycle - Phosphorus exists as a mineral and is not found in the atmosphere. Fungi and plant roots have a structure called mycorrhizae that are able to fix insoluble phosphates into useable phosphorus. Urine and decayed matter return phosphorus to the earth where it can be fixed in the plant. Phosphorus is needed for the backbone of DNA and for ATP manufacturing.

Skill 9.4   Analyze patterns of energy flow in the biosphere

Trophic levels are based on the feeding relationships that determine energy flow and chemical cycling.

Autotrophs are the primary producers of the ecosystem. **Producers** mainly consist of plants. **Primary consumers** are the next trophic level. The primary consumers are the herbivores that eat plants or algae. **Secondary consumers** are the carnivores that eat the primary consumers. **Tertiary consumers** eat the secondary consumer. These trophic levels may go higher depending on the ecosystem. **Decomposers** are consumers that feed off animal waste and dead organisms. This pathway of food transfer is known as the food chain. Most food chains are more elaborate, becoming food webs.

Whenever energy changes from one form to another, some of the original energy is lost as heat. For example, not all light striking a leaf is changed to chemical bond energy. Energy is also lost as the trophic levels progress from producer to tertiary consumer. The amount of energy that is transferred between trophic levels is called ecological efficiency. The visual of this energy flow is represented in a **pyramid of productivity**, where the greatest amount of usable energy is at the bottom (producers) with lower amounts of energy being available at the each level moving upward. This explains why vegetarian habits are encouraged to increase agricultural sustainability.

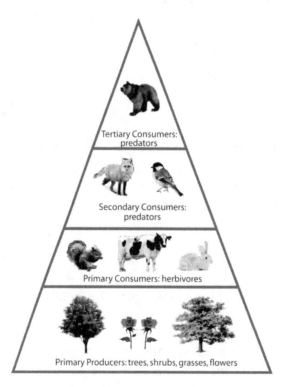

Depicted above, a **biomass pyramid** represents the total dry weight of organisms in each trophic level. A **pyramid of numbers** is a representation of the population size of each trophic level. The producers, being the most populous, are on the bottom of this pyramid with the tertiary consumers on the top with the fewest numbers.

Skill 9.5 **Evaluate factors that affect population composition, growth, size, and geographic distribution**

A **population** is a group of individuals of one species that live in the same general area. Many factors can affect the population size and its growth rate. Population size can depend on the total amount of life a habitat can support. This is the carrying capacity of the environment. Once the habitat runs out of food, water, shelter, or space, the carrying capacity decreases, and then stabilizes.

Limiting factors can affect population growth. As a population increases, the competition for resources is more intense, and the growth rate declines. This is a **density-dependent** growth factor. The carrying capacity can be determined by the density-dependent factor. **Density-independent factors** affect the individuals regardless of population size. The weather and climate are good examples. Too hot or too cold temperatures may kill many individuals from a population that has not reached its carrying capacity.

A zero population growth rate occurs when the birth and death rates are equal in a population. Exponential growth occurs when there is an abundance of resources and the growth rate is at its maximum, called the intrinsic rate of increase. This relationship can be graphically represented in a growth curve. An exponentially growing population begins with little change and then rapidly increases as seen in the J-curve below.

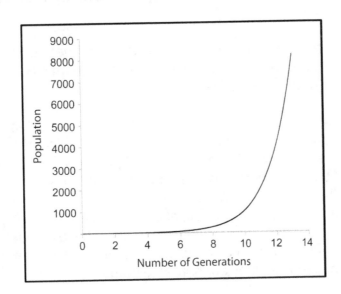

Logistic population growth incorporates the carrying capacity into the growth rate. As a population reaches the carrying capacity, the growth rate begins to slow down and level off as depicted in the S-curve below.

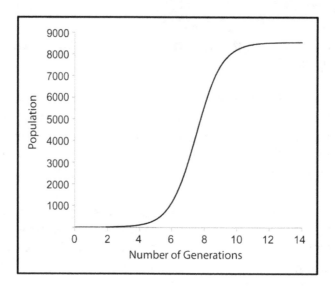

Many populations follow this model of population growth. Humans, however, are an exponentially growing population. Eventually, the carrying capacity of the Earth will be reached, and the growth rate will level off.

Skill 9.6   Distinguish between examples of competition, predation, and differing types of symbioses (e.g., parasitism, mutualism, commensalism)

There are many interactions that may occur between different species living together. Predation, parasitism, competition, commensalism, and mutualism are the different types of relationships populations have amongst each other.

Predation and **parasitism** result in a benefit for one species and a detriment for the other. Predation is when a predator eats its prey. The common conception of predation is of a carnivore consuming other animals. This is one form of predation. Although not always resulting in the death of the plant, herbivory is a form of predation. Some animals eat enough of a plant to cause death. Parasitism involves a predator that lives on or in their hosts, causing detrimental effects to the host. Insects and viruses living off and reproducing in their hosts is an example of parasitism. Many plants and animals have defenses against predators. Some plants have poisonous chemicals that will harm the predator if ingested and some animals are camouflaged so they are harder to detect.

Competition is when two or more species in a community use the same resources. Competition is usually detrimental to both populations. Competition is often difficult to find in nature because competition between two populations is not continuous. Either the weaker population will no longer exist, or one population will evolve to utilize other available resources.

Symbiosis is when two species live close together. Parasitism is one example of symbiosis described above. Another example of symbiosis is commensalisms. **Commensalism** occurs when one species benefits from the other without harmful effects. **Mutualism** is when both species benefit from the other. Species involved in mutualistic relationships must co-evolve to survive. As one species evolves, the other must as well if it is to be successful in life. The grouper and a species of shrimp live in a mutualistic relationship. The shrimp feed off parasites living on the grouper; thus the shrimp are fed and the grouper stays healthy. Many microorganisms are in mutualistic relationships.

Skill 9.7   Evaluate succession in communities

Succession is an orderly process of replacing a community that has been damaged or has begun where no life previously existed. Primary succession occurs where life never existed before, as in a flooded area or a new volcanic island. Secondary succession takes place in communities that were once flourishing but disturbed by some source, either man or nature, but not totally stripped. A climax community is a community that is established and flourishing. An example of succession is the gradual filling in of a lake community to become a climax grassland community.

Abiotic and biotic factors play a role in succession. **Biotic factors** are living things in an ecosystem; plants, animals, bacteria, fungi, etc. **Abiotic factors** are non-living aspects of an ecosystem; soil quality, rainfall, temperature, etc. Abiotic factors affect succession by way of the species that colonize the area. Certain species will or will not survive depending on the weather, climate, or soil makeup. Biotic factors such as inhibition of one species due to another may occur. This may be due to some form of competition between the species.

Since which species will become established depends upon soil type and climate, the stages of succession and the climax community for a given area are usually predictable.

Skill 9.8   Identify renewable and nonrenewable resources and compare management strategies for each, including environmental quality assessment and mitigation

The two categories of natural resources are renewable and nonrenewable. Renewable resources can be replaced as they are used, and as long as the rate of harvest does not exceed the rate of re-growth, use of these resources can be sustainable.. Examples of renewable resources are oxygen, wood, fresh water, and biomass. Nonrenewable resources are present in finite amounts or are used faster than they can be replaced in nature. Examples of nonrenewable resources are petroleum, coal, and natural gas.

Strategies for the management of renewable resources focus on balancing the immediate demand for resources with long-term sustainability. In addition, renewable resource management attempts to optimize the quality of the resources. For example, scientists may attempt to manage the amount of timber harvested from a forest, balancing the human need for wood with the future viability of the forest as a source of wood. Scientists attempt to increase timber production by fertilizing, manipulating trees genetically, and managing pests and density. Similar strategies exist for the management and optimization of water sources, air quality, and other plants and animals.

The main concerns in nonrenewable resource management are conservation, allocation, and environmental mitigation. Policy makers, corporations, and governments must determine how to use and distribute scare resources. Decision makers must balance the immediate demand for resources with the need for resources in the future. This determination is often the cause of conflict and disagreement. Finally, scientists attempt to minimize and mitigate the environmental damage caused by resource extraction, transportation, and use. Scientists devise methods of harvesting and using resources that do not unnecessarily impact the environment. After the extraction of resources from a location, scientists devise plans and methods to restore the environment to as close to its original state as possible.

Skill 9.9 Analyze the effects of resource availability on society

The availability of resources greatly affects the distribution of wealth and the patterns of population growth and development of society. In general, societies with large and varied supplies of natural resources are more prosperous than those lacking adequate resource supplies.

In addition, a sufficient supply of natural resources is necessary to sustain population growth. Countries that lack access to resources are usually less densely populated than those with abundant resources are. Finally, because resources are scarce, conflicts often arise over their distribution. Many disputes and armed conflicts between countries stem from tensions over resource allocation and availability.

Skill 9.10 Identify the local and global economic, aesthetic, and medical consequences of air, land, and water pollution and evaluate proposed solutions

All forms of pollution have both local and global economic, aesthetic, and medical consequences. Air, land, and water pollution directly and indirectly affect human health through the action of carcinogens, particulates, and other toxins that enter the air, water, and soil. Pollution negatively influences the local and global economy by increasing medical costs, increasing pollution treatment costs (e.g. water and soil clean up), and decreasing agricultural yields. Finally, all types of pollution decrease natural beauty and diminish enjoyment of nature and the outdoors.

Air pollution, possibly the most damaging form of pollution, has both local and global consequences. Air pollution results largely from the burning of fuels. Major sources of air pollution include transportation, industrial processes, heat and power generation, and the burning of solid waste. At the local level, air pollution negatively affects quality of life by increasing medical problems and decreasing comfort and enjoyment of outdoor activities. Air pollution can cause medical problems ranging from simple throat or eye irritations to asthma and lung cancer. Globally, air pollution threatens the Earth's ozone layer and excessive carbon dioxide emissions cause global warming. Ozone depletion increases the risk of sun related health problems (e.g. skin cancer) and global warming and the resulting climate changes threaten the Earth's ecological balance and biodiversity. Possible solutions to air pollution are government controls on fuel types, industry combustion standards, and the development and use of alternative, cleaner sources of energy.

Land pollution is the destruction of the Earth's surface resulting from improper industrial and urban waste disposal, damaging agricultural practices, mining and mineral exploitation. Land pollution greatly effects aesthetic appeal and human health. At the local level, improper waste disposal threatens the health of people living in the affected areas. Waste accumulation attracts pests and creates unsightly, dirty living conditions. Globally, damage and depletion of the soil by improper agricultural practices has great economic consequences. Overuse of pesticides and herbicides and depletion of soil nutrients causes long-term damage to the soil, leading to decreased crop yield in the future. Close regulation of waste disposal and agricultural practices is the most effective strategy for the prevention of land pollution.

Water pollution, perhaps the most prevalent form of pollution, greatly affects human health and the economy. The agricultural industry is the leading contributor to water pollution. Rain runoff carrying pesticides, herbicides, and fertilizer readily pollutes oceans, lakes, and rivers. Also carried into the oceans by wind and rain are large quantities of solid plastic waste, which drift in the ocean and causes damage to sea life and sea birds, who often mistake it for food and die after ingesting it.

Also contributing to water pollution are industrial effluents, sewage, and domestic waste. Water pollution negatively affects the economy because clean up and treatment of polluted water is very costly. In addition, consumption of polluted water causes health problems and increases associated medical costs. Finally, polluted water disrupts aquatic ecosystems, decreasing the availability of fish and other aquatic resources. Like air and land pollution, limiting water pollution requires strict governmental control. Proper oversight requires the implementation and enforcement of environmental standards regulating agricultural and industrial discharge into bodies of water.

Skill 9.11   Identify the potential local and global economic, aesthetic, and medical consequences of global warming and evaluate proposed solutions

The potential effects of global warming are far-reaching. The increasing concentration of carbon dioxide pollution threatens to increase average temperatures by 3 to 9 degrees by the end of the century. Global warming will affect weather patterns and increase sea levels. In addition, tropical diseases may also spread into new portions of the globe. Finally, global warming will disrupt ecosystems, causing species extinction and loss of species diversity. The possible solutions for controlling global warming involve the reduction of carbon dioxide in the atmosphere.

The effects of global warming on weather patterns and sea levels are likely to be large.. Warmer water occupies more volume than the same mass of cooler liquid water, so sea levels will rise. Melting glaciers will also cause sea levels to rise, flooding coastal areas and necessitating the relocation of coastal residents. Since heat fuels wind and water cycles, higher air temperatures also increase the likelihood of droughts, wildfires, heat waves, and intense rainstorms. In addition, higher ocean temperatures may increase the severity of hurricanes and tropical storms. Increased severe weather would negatively affect both local and global economies and cause a rise in heat and weather-related deaths and injuries.

Global warming will also increase the geographic range and virulence of tropical diseases because disease-carrying vectors (e.g. mosquitoes) will spread outside of their normal climatic zones. The spread of disease will have a potentially devastating effect on global health, impacting personal health and increasing the related economic costs of disease treatment and prevention. Other classes of organisms whose populations are typically controlled by low temperatures (such as fungi, microbes, and insects), may also expand their ranges as well as their annual active periods.

Finally, global warming disrupts ecosystems. Rising temperatures will cause the extinction of species that cannot adapt to the climatic changes. Loss of species diversity causes a dangerous imbalance in the ecosystem, especially when concurrent with shifts mentioned above that favor disease and insect populations.

The possible strategies for mitigating global warming involve methods for reducing atmospheric carbon levels. First, reducing energy use and finding and using alternative sources of energy (e.g. wind, solar, etc.) may slow global warming by reducing carbon emissions. In addition, carbon capture and sequestration are strategies for capturing carbon from major sources of carbon emissions (e.g. power plants) and storing it safely beneath the Earth's surface. Methane is another powerful greenhouse gas that can be controlled to mitigate global warming. Some legislation has been proposed to do so, but just as with atmospheric carbon pollution, efforts to control methane emissions are met with strong opposition from major industries.

Individual human organisms that wish to mitigate global warming, or "reduce their carbon footprint," can choose from among a variety of strategies: buy solar or wind energy; use bicycles, public transportation, or fuel-efficient cars; eat less meat and more locally grown food; reduce energy waste and loss from the home; invest in retirement funds that support non-carbon fuel and sustainable companies; and support bipartisan efforts in government that seek to address climate change issues.

Skill 9.12   Analyze the local and global consequences of loss of biodiversity

The current species extinction rate is an estimated 100 times greater than the pre-human rate. This rapid loss of biodiversity has potentially grave consequences beyond the typical aesthetic and ethical concerns. Ecological disruption caused by the loss of biodiversity can negatively affect human health and hinder medical treatment and research.

Species of plants, animals, and microorganisms interact to complete many important ecological tasks such as gas exchange, water filtration, soil fertilization, and temperature and precipitation regulation. The loss of only a few species to extinction can disrupt the fine balance that drives these important processes. Even species that are functionally redundant are important because they serve as buffers against environmental change. Environmental change may lessen the productivity of one species, and the presence of another species that can fulfill the same ecological function allows the ecosystem to continue without disruption. However, when environmental change affects many species, as is currently the case with climate change, it is more likely there will be disruption. Disruption of ecosystem function due to loss of biodiversity can negatively affect food production and human health. Increased pollution and decreased soil quality are but two examples.

A lack of biodiversity can also induce the emergence and spread of infectious disease. Loss of diversity can decrease competition for disease-vector species allowing population growth and increased disease-spreading capability. For example, the extinction of the natural competitors and predators of white footed mice in North American forests led to an increase in the population of this Lyme disease-carrying species. As a result, incidence of Lyme disease skyrocketed.

Finally, loss of biodiversity can negatively affect human health by eliminating potential medical treatments and hampering medical research. Many pharmaceutical drugs come from natural sources and the premature loss of species eliminates the possibility of new drug discovery. In addition, studying different species helps scientists understand physiology and disease mechanisms. In other words, medical research suffers when species become extinct.

Skill 9.13   Characterize ecosystems unique to Florida (i.e., terrestrial, marine, freshwater) and identify indicator species of each

An ecosystem is a community of plants and animals that live together. Ecosystems found in South Florida include flatwoods, coral reefs, dunes, marshes, swamps, hammocks, and mangroves. Florida is also the home of the well-known Everglades National Park.

Flatwoods

The most extensive terrestrial ecosystem in Florida is the pine flatwoods. This evolved under frequent fire, seasonal drought, and flooded soil conditions.

The pine flatwoods can be divided into two groups: the North Florida flatwoods which are typically open woodlands dominated by pine trees, and the South Florida flatwoods which are typically savannas. The flatwoods consist primarily of various pine trees and an understory of shrubs such as the saw palmetto, wax myrtle, wildflowers, ferns, and blueberries. Of the underbrush, the Chapman's rhododendron is currently listed as threatened/endangered. Plants that grow in the pinelands must be resistant to fire because pinelands are maintained by fire. Fires are beneficial to the pines because young pine seedlings require lots of sunlight to survive, and the fires destroy hardwood competitors. When fires occur, hardwood seedlings and other understory plants are affected, but the thick bark of the pine resists fire damage. Wildlife found here include deer, squirrels, bobcats, skunks, opossums, raccoons, birds, snakes, and tortoises.

Coral Reefs

Southern Florida's ecosystem contains the only living continuous coral reef system adjacent to the continental U.S. Over 30 different kinds of corals, including the Star and Staghorn Corals are found in Florida waters. Florida waters are the principal nursery for the commercial and sport fisheries in Florida.

Dunes

Dunes are created by wind, but are held in place by grasses that trap sand grains as they are being moved across the beach. Dunes stabilized by grasses protect the coast against winds and pounding waves. Florida beaches are important nesting sites for sea turtles and shorebirds. A loss of beach habitat to real estate development, erosion, and rising sea level has caused a decline in the nesting shorebird and sea turtle populations.

Freshwater Marshes

Freshwater marshes are generally wetlands with an open expanse of grasses and other grass-like plants. They have standing water for much of the year and act as natural filters, slowing down the water's movement and allowing the settling of particles.

Animals found in the marsh can include fish, mollusks, shrimp, frogs, snakes, alligators, and the threatened Florida panther.

Freshwater Swamps

Freshwater swamps are wet, wooded areas where standing water occurs for at least part of the year. The freshwater swamps may have cypress trees, bay trees or hardwoods. Other plants found in swamps include epiphytes ("air plants") growing on trees, vines, and ferns. Wood storks, herons, otters, black bear, and the Florida panther are only a few of the animals that find food, homes, and nesting sites in Florida's swamps.

Hardwood Hammocks

Hardwood hammocks are small areas of hardwood trees that can grow on natural rises of land. In Florida, hammocks occur in marshes, pinelands, and mangrove swamps. Hammocks may contain many different species of trees such as the sabal palm, live oak, red maple, mahogany, gumbo limbo and coco plum.

Wildlife in hammocks can include tree snails, raccoons, opossums, birds, snakes, lizards, tree frogs, as well as large mammals.

Mangroves

Three species of mangroves are found in Florida: the red mangrove, black mangrove, and white mangrove. Red mangroves grow along the water's edge, followed by black, while white mangroves grow mostly inland. Mangroves grow in saltwater and in areas frequently flooded by saltwater. Mangroves provide protected habitat, breeding grounds, and nursery areas to many land and marine animals. Mangroves also provide shoreline protection from wind, waves, and erosion. Of special note is the presence of the beloved manatee, who can be found here, especially where it's preferred food source, seagrass, resides.

Competency 10   Knowledge of evolutionary mechanisms

Skill 10.1   Compare evolution by natural selection with other theories (e.g., Lamarck, Darwin)

Charles Darwin proposed a mechanism for his theory of evolution, which he termed natural selection. Natural selection describes the process by which favorable traits accumulate in a population, changing the population's genetic make-up over time. Darwin theorized that all individual organisms, even those of the same species, are different and those individuals that happen to possess traits favorable for survival would produce more offspring. Thus, in the next generation, the number of individuals with the favorable trait increases and the process continues. Darwin, in contrast to other evolutionary scientists, did not believe that traits acquired during an organism's lifetime (e.g. increased musculature) or the desires and needs of the organism affected evolution of populations. For example, Darwin argued that the evolution of long trunks in elephants resulted from environmental conditions that favored those elephants that possessed longer trunks. The individual elephants did not stretch their trunks to reach food or water and pass on the new, longer trunks to their offspring.

Jean Baptiste Lamarck proposed an alternative mechanism of evolution. Lamarck believed individual organisms developed traits in response to changing environmental conditions and passed on these new, favorable traits to their offspring. For example, Lamarck argued that the trunks of individual elephants lengthen as a result of stretching for scarce food and water, and elephants pass on the longer trunks to their offspring. Thus, in contrast to Darwin's relatively random natural selection, Lamarck believed the mechanism of evolution followed a predetermined plan and depended on the desires and needs of individual organisms.

Skill 10.2   Analyze the classical species concept and its limitations

The most commonly used species concept is the **Biological Species Concept (BSC)**. This states that a species is a reproductive community of populations that occupy a specific niche in nature. It focuses on reproductive isolation of populations as the primary criterion for recognition of species status. The biological species concept does not apply to organisms that are completely asexual in their reproduction, fossil organisms, or distinctive populations that hybridize.

Reproductive isolation is caused by any factor that impedes two species from producing viable, fertile hybrids. Reproductive barriers can be categorized as **prezygotic** (premating) or **postzygotic** (postmating).

The prezygotic barriers are as follows:

1. Habitat isolation – species occupy different habitats in the same territory.
2. Temporal isolation – populations reaching sexual maturity/flowering at different times of the year.
3. Ethological isolation – behavioral differences that reduce or prevent interbreeding between individuals of different species (including pheromones and other attractants).
4. Mechanical isolation – structural differences that make gamete transfer difficult or impossible.
5. Gametic isolation – male and female gametes do not attract each other; no fertilization.

The postzygotic barriers are as follows:

1. Hybrid inviability – hybrids die before sexual maturity.

2. Hybrid sterility – disrupts gamete formation; no normal sex cells.
3. Hybrid breakdown – reduces viability or fertility in progeny of the F_2 backcross.

Though useful for classification, the species distinction is not always clear-cut. Animals may exist as separate, reproductively isolated groups in nature, when it is possible that prezygotic barriers to their reproduction might be removed. Scientists sometimes refer to subspecies, and recognize that reproductive barriers change over time.

Skill 10.3 **Compare systems of classification (e.g., classical taxonomy, phenetics, cladistics)**

Taxonomy is the science of naming, classifying, and describing living organisms. Different systems of classification group organisms based on different characteristics and criteria. Classical taxonomy is notoriously subjective, classifying and naming organisms based on easily observable characteristics and ancestral heritage. Alternative systems of classification, including phenetics, cladistics, and molecular taxonomy, attempt to remove the subjective nature of classification.

Classical taxonomy classifies organisms based on shared and ancestral characteristics. While this approach is more accurate than classifying organisms based strictly on observable characteristics, it is relatively subjective because the characteristics used to classify may or may not be important from an evolutionary perspective.

Phenetics, or numerical taxonomy, classifies organisms based on overall similarity. Phenetic analysis determines overall similarity by quantifying the number of shared characteristics (e.g. morphological, anatomical, behavioral) between species. While the phenetic approach to classification eliminates some of the subjectivity of classical taxonomy, it does not consider evolutionary heritage.

Cladistics is a system of classification that determines evolutionary relationships between organisms based on shared derived traits.

When analyzing the evolutionary relationship between species of animals in cladistics, observable traits shared by species are only important if they derive from a common ancestor. In other words, two groups of animals may share many characteristics but may not be closely related because the characteristics they share derive from an ancestor common to all animals.

Finally, molecular taxonomy classifies organisms based on evolutionary relationships determined by DNA and protein composition. Molecular technology allows for objective analysis of relationships, independent of arbitrary observation of traits and characteristics. Protein structure, percentage of shared DNA, and signature DNA sequences can reveal evolutionary relationships between species of organisms.

Skill 10.4   Apply a taxonomic key to a set of objects

A taxonomic key, also known as a dichotomous key, allows the identification of objects and the classification of a set of organisms into species-level groups. Taxonomic keys divide objects or organisms progressively by offering two (or more) options for each observed characteristic. The answer to each successive question in the key divides the objects or organisms into two (or more) mutually exclusive groups and determines the next step in the classification process. The process continues until the objects or organisms (or their species/group) are identified.

For example, the following is a basic taxonomic key used to divide and identify five everyday objects, a rubber eraser, a nail, a rubber band, a writing pen, and a penny.

1. Contains metal: go to 2
1. Does not contain metal: go to 3

2. Flat, disc-shaped: penny
2. Not flat or disc-shaped: nail

3. Does not contain rubber: writing pen
3. Contains rubber: go to 4

4. Ring-shaped: rubber band
4. Not ring-shaped: rubber eraser

Note that the key positively identifies the objects by separating them into two mutually exclusive groups at each step.

Scientists design taxonomic keys, similar in structure to the preceding example, to identify the species of observed organisms. For example, field biologists may use a taxonomic key that identifies oak trees based on leaf structure to assign oak trees observed in the field to their appropriate species.

Skill 10.5   Analyze variation within a species and its relationship to changes along an environmental cline

In addition to variations between species, individual organisms within species vary. Members of a species, generally defined as a group of organisms able to produce viable offspring, often vary in many characteristics because of evolutionary adaptations to environmental conditions that differ from one area to another. Polymorphism is the coexistence of distinct types within a species population. Human blood type is an example of a polymorphic trait. Species populations in different geographic locations also display differences in characteristics. An environmental cline is a pattern of gradual change in a characteristic(s) over the geographical range of a species. For example, the size of mammals of many northern hemisphere species gradually increases from south to north along the climatic gradient. This is assumed to be an adaptation to better preserve body heat, since the smaller surface area: volume ratio of the larger body allows this. Conversely, a smaller body cools more efficiently in hot climates.

Depending on the amount of gene flow, transfer of genes between geographically separated populations, environmental clines can be smooth or stepped. When gene flow is high, the cline is smooth or gradual. When gene flow is low, the cline is stepped or abrupt. A smooth cline, representing a gradual change in a characteristic, exists when the geographic barrier to reproduction between populations is slight. For example, changes resulting from a temperature gradient produce a smooth cline because the reproductive barrier is minimal and, thus, gene flow is high. A stepped cline, representing an abrupt change in a characteristic, exists when the geographic barrier to reproduction is significant. For example, because the gene flow between populations separated by deserts or mountains is low, the change in an observed characteristic is abrupt.

Skill 10.6   Identify factors affecting speciation and evolution in general (e.g., mutation, recombination, isolation, sexual reproduction and selection, genetic drift, plate tectonics and geographic distribution)

1. Heritable variation
2. Mutations
3. Recombination
4. Natural selection
5. Sexual selection

Heritable variation is responsible for the individuality of organisms. An individual's phenotype is based on inherited genotype and the surrounding environment. For example, people can alter their phenotypes by lifting weight or diet and exercise.

Variation is generated by mutation and sexual recombination. **Mutations** may be errors in replication or a spontaneous rearrangement of one or more segments of DNA.

Mutations contribute a minimal amount of variation in a population. It is the unique **recombination** of existing alleles that causes the majority of genetic differences.

Recombination is caused by the crossing over of the parents' genes during meiosis. This results in a unique offspring. With all the possible mating combinations in the world, it is obvious how sexual reproduction is the primary cause of genetic variation.

Natural selection is based on the survival of certain traits in a population through the course of time. The phrase "survival of the fittest," is often associated with natural selection. Fitness is the contribution an individual makes to the gene pool of the next generation.

Natural selection acts on phenotypes. An organism's phenotype is constantly exposed to its environment. Based on an organism's phenotype, selection indirectly adapts a population to its environment by maintaining favorable genotypes in the gene pool.

There are three modes of natural selection. Stabilizing selection favors the more common phenotypes, directional selection shifts the frequency of phenotypes in one direction, and diversifying selection occurs when individuals on both extremes of the phenotypic range are favored.

Sexual selection leads to the secondary sex characteristics between male and females. Animals that use mating behaviors may be successful or unsuccessful. An animal that lacks attractive plumage or has a weak mating call will not attract the female, thereby eventually limiting that gene in the gene pool. Mechanical isolation, where sex organs do not fit the female, has an obvious disadvantage.

Skill 10.7   Evaluate the role of mutation, recombination, isolation, sexual reproduction and selection, genetic drift, and plate tectonics and geographic distribution on evolution

Different molecular and environmental processes and conditions drive the evolution of populations. The various mechanisms of evolution either introduce new genetic variation or alter the frequency of existing variation.

Mutations, random changes in nucleotide sequence, are a basic mechanism of evolution. Mutations in DNA result from copying errors during cell division, exposure to radiation and chemicals, and interaction with viruses. Simple point mutations, deletions, or insertions can alter the function or expression of existing genes but do not contribute greatly to evolution. On the other hand, gene duplication, the duplication of an entire gene, often leads to the creation of new genes that may contribute to the evolution of a species. Because gene duplication results in two copies of the same gene, the extra copy is free to mutate and develop without the selective pressure experienced by mutated single-copy genes. Gene duplication and subsequent mutation often lead to the creation of new genes. When new genes resulting from mutations lend the mutated organism a reproductive advantage relative to environmental conditions, natural selection and evolution can occur.

Recombination is the exchange of DNA between a pair of chromosomes during meiosis. Recombination does not introduce new genes into a population, but does affect the expression of genes and the combination of traits expressed by individuals. Thus, recombination increases the genetic diversity of populations and contributes to evolution by creating new combinations of genes that nature selects for or against.

Isolation is the separation of members of a species by environmental barriers that the organisms cannot cross. Environmental change, either gradual or sudden, often results in isolation. An example of gradual isolation is the formation of a mountain range or dessert between members of a species. An example of sudden isolation is the separation of species members by a flood or earthquake. Isolation leads to evolution because the separated groups cannot reproduce together and differences arise. In addition, because the environment of each group is different, the groups adapt and evolve differently. Extended isolation can lead to speciation, the development of new species.

Sexual reproduction and selection contributes to evolution by consolidating genetic mutations and creating new combinations of genes. Genetic recombination during sexual reproduction, as previously discussed, introduces new combinations of traits and patterns of gene expression. Consolidation of favorable mutations through sexual reproduction speeds the processes of evolution and natural selection. On the other hand, consolidation of deleterious mutations creates completely unfit individuals that are readily eliminated from the population.

Genetic drift is, along with natural selection, one of the two main mechanisms of evolution. Genetic drift refers to the chance deviation in the frequency of alleles (traits) resulting from the randomness of zygote formation and selection. Because only a small percentage of all possible zygotes become mature adults, parents do not necessarily pass all of their alleles on to their offspring. Genetic drift is particularly important in small populations because chance deviations in allelic frequency can quickly alter the genotypic make-up of the population. In extreme cases, certain alleles may completely disappear from the gene pool. Genetic drift is particularly influential when environmental events and conditions produce small, isolated populations.

Plate tectonics is the theory that the Earth's surface consists of large plates. Movement and shifting of the plates dictate the location of continents, formation of mountains and seas, and volcanic and earthquake activity. Such contributions to environmental conditions influence the evolution of species. For example, tectonic activity resulting in mountain formation or continent separation can cause genetic isolation. In addition, the geographic distribution of species is indicative of evolutionary history and related tectonic activity.

Skill 10.8   Compare the concepts of punctuated equilibrium and gradualism

There are two theories on the rate of evolution. **Gradualism** is the theory that minor evolutionary changes occur at a regular rate. Darwin's book, "On the Origin of Species," is based on this theory of gradualism. Charles Darwin was born in 1809 and spent 5 years in his twenties on a ship called the *Beagle*. Of all the locations the *Beagle* sailed to, it was the Galapagos Islands that infatuated Darwin. There he collected 13 species of finches that were quite similar. He could not accurately determine whether these finches were of the same species. He later learned these finches were in fact separate species. Darwin began to hypothesize that new species arose from its ancestors by the gradual collection of adaptations to a different environment. Darwin's most popular hypothesis is on the beak size of Galapagos finches. He theorized that the finches' beak sizes evolved to accommodate different food sources. Many people did not believe in Darwin's theories until recent field studies proved successful. Although Darwin believed the origin of species was gradual, he was bewildered by the gaps in fossil records of living organisms. **Punctuated equilibrium** is the model of evolution that states that organismal form diverges and species form rapidly over relatively short periods of geological history, and then progress through long stages of stasis with little or no change. Punctuationalists use fossil records to support their claim. It is probable that both theories are correct, depending on the particular lineage studied.

Skill 10.9 Distinguish between examples of evidences for evolutionary theory (e.g., biochemical, morphological, embryological, paleontological)

The wide range of evidence of evolution provides information on the natural processes by which the variety of life on earth developed.

1. Paleontology:

Paleontology is the study of past life based on fossil records and their relation to different geologic time periods.

When organisms die, they often decompose quickly or are consumed by scavengers, leaving no evidence of their existence. However, occasionally some organisms are preserved. The remains or traces of the organisms from a past geological age embedded in rocks by natural processes are called fossils. They are very important for the understanding the evolutionary history of life on earth as they provide evidence of evolution and detailed information on the ancestry of organisms.

Petrification is the process by which a dead animal gets fossilized. For this to happen, a dead organism must be buried quickly to avoid weathering and decomposition. When the organism is buried, the organic matter decays. The mineral salts from the mud (in which the organism is buried) will infiltrate into the bones and gradually fill up the pores. The bones will harden and be preserved as fossils. If dead organisms are covered by wind- blown sand and if the sand is subsequently turned into mud by heavy rain or floods, the same process of mineral infiltration may occur. Besides petrification, the organisms may be well preserved in ice, in hardened resin of coniferous trees (amber), in tar, in anaerobic acidic peat. Fossilization can sometimes be a trace, an impression of a form – e.g., leaves and footprints.

From the horizontal layers of sedimentary rocks (these are formed by silt or mud on top of each other) called strata and each layer consists of fossils. The oldest layer is the one at the bottom of the pile and the fossils found in this layer are the oldest; this is how the paleontologists determine the relative ages of these fossils.

Some organisms appear in some layers only indicating that they lived only during that period and became extinct. A succession of animals and plants can also be seen in fossil records, which supports the theory that organisms tend to progressively increase in complexity.

According to fossil records, some modern species of plants and animals are found to be almost identical to the species that lived in ancient geological ages. They are existing species of ancient lineage that have remained unchanged morphologically and maybe physiologically as well. Hence, they're called "living fossils". Some examples of living fossils are tuatara, nautilus, horseshoe crab, gingko and metasequoia.

Fossils are the key to understanding biological history. They are the preserved remnants left by an organism that lived in the past. Scientists have established the geological time scale to determine the age of a fossil. The geological time scale is broken down into four eras: the Precambrian, Paleozoic, Mesozoic, and Cenozoic. The eras are further broken down into periods that represent a distinct age in the history of Earth and its life. Scientists use rock layers called strata to date fossils. The older layers of rock are at the bottom. This allows scientists to correlate the rock layers with the era they date back to. Radiometric dating is a more precise method of dating fossils. Rocks and fossils contain isotopes of elements accumulated over time. The isotope's half-life is used to date older fossils by determining the amount of isotope remaining and comparing it to the half-life.

Dating fossils is helpful to construct an evolutionary tree. Scientists can arrange the succession of animals based on their fossil record. The fossils of an animal's ancestors can be dated and placed on its evolutionary tree. For example, the branched evolution of horses shows the progression of the modern horse's small ancestors to forms which are larger, have a reduced number of toes, and have teeth modified for grazing.

2. Anatomy:

Comparative anatomical studies reveal that some structural features are basically similar – e.g., flowers generally have sepals, petals, stigma, style and ovaries but the size, color, number of petals, sepals etc., may differ from species to species.

The degree of resemblance between two organisms indicates how closely they are related in evolution.

- Groups with little in common are supposed to have diverged from a common ancestor much earlier in geological history than groups which have more in common
- To decide how closely two organisms are, anatomists look for the structures which may serve a different purpose in the adult, but are basically similar (homologous)
- In cases where similar structures serve different functions in adults, it is important to trace their origin and embryonic development

When a group of organisms share a homologous structure, which is specialized, to perform a variety of functions in order to adapt to different environmental conditions are called adaptive radiation. The gradual spreading of organisms with adaptive radiation is known as divergent evolution.

Examples of divergent evolution are – pentadactyl limb and insect mouth parts

Under similar environmental conditions, fundamentally different structures in different groups of organisms may undergo modifications to serve similar functions. This is called convergent evolution. The structures, which have no close phylogenetic links but showing adaptation to perform the same functions, are called analogous. Examples are – wings of bats, bird and insects, jointed legs of insects and vertebrates, eyes of vertebrates and cephalopods.

Vestigial organs: Organs that are smaller and simpler in structure than corresponding parts in the ancestral species are called vestigial organs. They are usually degenerated or underdeveloped. These were functional in ancestral species but have become nonfunctional, e.g., vestigial hind limbs of whales, vestigial leaves of some xerophytes, vestigial wings of flightless birds like ostriches, etc.

3. *Geographical distribution:*

- Continental distribution: All organisms are adapted to their environment to a greater or lesser extent. It is generally assumed that the same type of species would be found in a similar habitat in a similar geographic area. Examples: Africa has short tailed (old world) monkeys, elephants, lions and giraffes. South America has long-tailed monkeys, pumas, jaguars and llamas.
- Evidence for migration and isolation: The fossil record shows that evolution of camels started in North America, from which they migrated across the Bering Strait into Asia and Africa and through the Isthmus of Panama into South America.
- Continental drift: Fossils of the ancient amphibians, arthropods and ferns are found in South America, Africa, India, Australia and Antarctica which can be dated to the Paleozoic Era, at which time they were all in a single landmass called Gondwana.
- Oceanic Island distribution: Most small isolated islands only have native species.

Plant life in Hawaii could have arrived as airborne spores or as seeds in the droppings of birds. A few large mammals present in remote islands were brought by human settlers.

4. *Evidence from comparative embryology:*

Comparative embryology shows how embryos start off looking the same. As they develop their similarities slowly decrease until they take the form of their particular class.

Example: Adult vertebrates are diverse, yet their embryos are quite similar at very early stages. Fishlike structures still form in early embryos of reptiles, birds and mammals. In fish embryos, a two-chambered heart, some veins, and parts of arteries develop and persist in adult fishes. The same structures form early in human embryos but do not persist as in adults.

5. *Physiology and Biochemistry:*

Evolution of widely distributed proteins and molecules: All organisms make use of DNA and/or RNA. ATP is the metabolic currency. Genetic code is same for almost every organism. A piece of RNA in a bacterium cell codes for the same protein as in a human cell.

Comparison of the DNA sequence allows organisms to be grouped by sequence similarity, and the resulting phylogenetic trees are typically consistent with traditional taxonomy, and are often used to strengthen or correct taxonomic classifications. DNA sequence comparison is considered strong enough to be used to correct erroneous assumptions in the phylogenetic tree in cases where other evidence is missing. The sequence of the 168rRNA gene, a vital gene encoding a part of the ribosome was used to find the broad phylogenetic relationships between all life.

The proteomic evidence also supports the universal ancestry of life. Vital proteins such as ribosome, DNA polymerase, and RNA polymerase are found in the most primitive bacteria to the most complex mammals.

Since metabolic processes do not leave fossils, research into the evolution of the basic cellular processes is done largely by comparison of existing organisms.
There are other evidences like Complex Iteration and Speciation.

Skill 10.10 Analyze aspects of modern theories on the origin of life on Earth

The hypothesis that life developed on Earth from nonliving materials is the most widely accepted theory on the origin of life. The transformation from nonliving materials to life had four stages. The first stage was the nonliving (abiotic) synthesis of small monomers such as amino acids and nucleotides. In the second stage, these monomers combine to form polymers, such as proteins and nucleic acids. The third stage is the accumulation of these polymers into droplets called protobionts. The last stage is the origin of heredity, with RNA as the first genetic material.

The first stage of this theory was hypothesized in the 1920s. A. I. Oparin and J. B. S. Haldane were the first to theorize that the primitive atmosphere was a reducing atmosphere with no oxygen present. The gases were rich in hydrogen, methane, water and ammonia. In the 1950s, Stanley Miller proved Oparin's theory in the laboratory by combining the above gases. When given an electrical spark, he was able to synthesize simple amino acids. It is commonly accepted that amino acids appeared before DNA.

Other laboratory experiments have supported the other stages in the origin of life theory could have happened.

Other scientists believe simpler hereditary systems originated before nucleic acids. In 1991, Julius Rebek was able to synthesize a simple organic molecule that replicates itself. According to his theory, this simple molecule may be the precursor of RNA.

Skill 10.11   Recognize general evolutionary trends as they relate to major taxa

An evolutionary trend is a directional change within a single taxa or across several taxa. In other words, either the members of a single taxa or several different taxa evolve in the same general direction. Scientists have observed several trends in the evolution of the major taxa. Defining trends in evolution, however, is very difficult because there are always exceptions.

Scientists have identified several evolutionary trends that appear in different lineages and taxa. Most types of animals have trended toward increased body size and increased brain size in relation to body size. In addition, chordates, annelids (segmented worms), and arthropods (crustaceans and insects) display an evolutionary trend toward the organization of neurons within a head, or cephalization. Cephalization is a good example of a general evolutionary trend that has many exceptions. Many animal lineages (e.g. starfish) have not developed any sort of head-like structure and some internal parasites have evolved in the reverse direction, losing the heads they once had. Finally, increased complexity is a global evolutionary trend. While many species remain relatively simple in structure, (e.g. bacteria) the number of complex species has greatly increased over time. It is interesting to note, however, that the rise of complex organisms has not diminished the importance of simple organisms to ecological stability.
Another major evolutionary trend has been the movement from aquatic to terrestrial environments. Early life forms depended on an aqueous environment to survive. Multiple adaptations to allow for life and reproduction on land have occurred in plants and animals, including the evolution of lungs, shelled eggs, pollen, and protective coverings to preserve internal water such as scales, skin, and plant waxes.

In addition to cross-lineage evolutionary trends, individual lineages and taxa often display unique directional changes. In other words, each taxa has its own evolutionary trends. For example, many types of trees and plants have developed remarkably similar patterns of leaf structure and plant flowers have trended away from radial symmetry and toward bilateral symmetry.

Apply the Hardy-Weinberg formula and identify the assumptions upon which it is based

Evolution currently is defined as a change in genotype over time. Gene frequencies shift and change from generation to generation. Populations evolve, not individuals. The **Hardy-Weinberg** theory of gene equilibrium is a mathematical prediction to show shifting gene patterns. Let's use the letter "A" to represent the dominant condition of normal skin pigment. "a" would represent the recessive condition of albinism. In a population, there are three possible genotypes; AA, Aa and aa. AA and Aa would have normal skin pigment and only aa would be albinos. According to the Hardy-Weinberg law, there are five requirements to keep a gene frequency stable, leading to no evolution:

1. There is no mutation in the population.
2. There are no selection pressures; one gene is not more desirable in the environment.
3. There is no mating preference; mating is random.
4. The population is isolated; there is no immigration or emigration.
5. The population is large (mathematical probability is more correct with a large sample).

The above conditions are extremely difficult to meet. If these five conditions are not met, then gene frequency can shift, leading to evolution. Let's say in a population, 75% of the population has normal skin pigment (AA and Aa) and 25% are albino (aa). Using the following formula, we can determine the frequency of the A allele and the "a" allele in a population.

This can be used over generations to determine if evolution is occurring. The formula is: $1 = p^2 + 2pq + q^2$; where 1 is the total population. p^2 is the number of AA individuals, $2pq$ is the number of Aa individuals, and q^2 is the number of aa individuals.

Since you cannot tell by looking if an individual is AA or Aa, you must use the aa individuals to find that frequency first. As stated above aa was 25% of the population. Since $aa = q^2$, we can determine the value of q (or a) by finding the square root of 0.25, which is 0.5. Therefore, 0.5 of the population has the "a" gene. In order to find the value for p, use the following formula: $1 = p + q$. This would make the value of $p = 0.5$. The gene pool is all the alleles at all gene loci in all individuals of a population. The Hardy-Weinberg theorem describes the gene pool in a non-evolving population. It states that the frequencies of alleles and genotypes in a population's gene pool are random unless acted on by something other than sexual recombination.

Now, to find the number of *AA*, plug it into the first formula;

$$AA = p^2 = 0.5 \times 0.5 = 0.25$$
$$Aa = 2pq = 2(0.5 \times 0.5) = 0.5$$
$$aa = q^2 = 0.5 \times 0.5 = 0.25$$

Any problem you may have with Hardy Weinberg will have an obvious squared number. The square of that number will be the frequency of the recessive gene, and you can figure anything else out knowing the formula and the frequency of *q*!

When frequencies vary from the Hardy Weinberg equilibrium, the population is said to be evolving. The change to the gene pool is on such a small scale that it is called microevolution. Certain factors increase the chances of variability in a population, thus leading to evolution. Items that increase variability include mutations, sexual reproduction, immigration, large population, and variation in geographic local. Changes that decrease variation would be natural selection, emigration, small population, and random mating.

Sample Test

Directions: Read each item and select the best response.

1. **A student designed a science project testing the effects of light and water on plant growth. You would recommend that she** *(Average Rigor) (Skill 1.12)*

   A. manipulate the temperature as well.

   B. also alter the pH of the water as another variable.

   C. omit either water or light as a variable.

   D. also alter the light concentration as another variable.

2. **Identify the control in the following experiment. A student had four plants grown under the following conditions and was measuring photosynthetic rate by measuring mass. 2 plants in 50% light and 2 plants in 100% light.** *(Rigorous) (Skill 1.16)*

   A. plants grown with no added nutrients

   B. plants grown in the dark

   C plants in 100% light

   D. plants in 50% light

3. **In an experiment measuring the growth of bacteria at different temperatures, identify the independent variable.** *(Rigorous) (Skill 1.9)*

   A. growth of number of colonies

   B. temperature

   C. type of bacteria used

   D. light intensity

4. **A scientific theory** *(Average Rigor)(Skill 1.8)*

   A. proves scientific accuracy.

   B. is never rejected.

   C. results in a medical breakthrough.

   D. may be altered at a later time.

5. Which is the correct order of methodology?

1) testing revised explanation
2) setting up a controlled experiment to test explanation
3) drawing a conclusion
4) suggesting an explanation for observations
5) compare observed results to hypothesized results
(Rigorous)(Skill 1.8)

A. 4, 2, 3, 1, 5

B. 3, 1, 4, 2, 5

C. 4, 2, 5, 1, 3

D. 2, 5, 4, 1, 3

6. Given a choice, which is the most desirable method of heating a substance in the lab? *(Easy) (Skill 2.6)*

A. alcohol burner

B. gas burner

C. Bunsen burner

D. hot plate

7. Biological waste should be disposed of *(Easy) (Skill 2.6)*

A. in the trash can.

B. under a fume hood.

C. in the broken glass box.

D. in an autoclavable biohazard bag.

8. Chemicals should be stored *(Easy) (Skill 2.6)*

A. in a cool dark room.

B. in a dark room.

C. according to their reactivity with other substances.

D. in a double locked room.

9. Given the choice of lab activities, which would you omit? *(Rigorous) (Skill 1.3)*

A. a genetics experiment tracking the fur color of mice

B. dissecting a preserved fetal pig

C. a lab relating temperature to respiration rate using live goldfish.

D. pithing a frog to see the action of circulation

10. Who should be notified in the case of a serious chemical spill?

   I. the custodian
   II. The fire department
   III. the chemistry teacher
   IV. the administration
   (Easy) (Skill 2.6)

   A. I

   B. II

   C. II and III

   D. II and IV

11. The "Right to Know" law states *(Easy) (Skill 2.6)*

   A. the inventory of toxic chemicals checked against the "Substance List" be available.

   B. that students are to be informed on alternatives to dissection.

   C. that science teachers are to be informed of student allergies.

   D. that students are to be informed of infectious microorganisms used in lab.

12. In which situation would a science teacher be liable? *(Average Rigor) (Skill 2.6)*

   A. a teacher leaves to receive an emergency phone call and a student slips and falls.

   B. a student removes their goggles and gets dissection fluid in their eye.

   C. a faulty gas line results in a fire.

   D. a students cuts themselves with a scalpel.

13. Which statement best defines negligence? *(Average Rigor) (Skill 2.6)*

   A. failure to give oral instructions for those with reading disabilities

   B. failure to exercise ordinary care

   C. inability to supervise a large group of students.

   D. reasonable anticipation that an event may occur

14. **Which item should always be used when using chemicals with noxious vapors?**
(Easy) (Skill 2.6)

   A. eye protection

   B. face shield

   C. fume hood

   D. lab apron

15. **Identify the correct sequence of the organization of living things.**
(Average Rigor) (Skill 10.3)

   A. cell – organelle – organ – tissue – organ system – organism

   B. cell – tissue – organ – organelle – organ system – organism

   C. organelle – cell – tissue – organ – organ system – organism

   D. organ system – tissue – organelle – cell – organism – organ

16. **Which is not a characteristic of living things?**
(Average Rigor) (Skill 4.2)

   A. movement

   B. cellular structure

   C. metabolism

   D. reproduction

17. **Which kingdom is comprised of organisms made of one cell with no nuclear membrane?**
(Average Rigor) (Skill 4.2)

   A. Monera

   B. Protista

   C. Fungi

   D. Algae

18. **Potassium chloride atoms are joined bya(n)**
(Rigorous) (Skill 3.1)

   A. non polar covalent bond

   B. polar covalent bond

   C. ionic bond

   D. hydrogen bond

19. Which of the following is a monomer? *(Rigorous) (Skill 3.1)*

A. RNA

B. glycogen

C. DNA

D. amino acid

20. Which of the following are properties of water?

I. High specific heat
II. Strong ionic bonds
III. Good solvent
IV. High freezing point
(Rigorous) (Skill 3.1)

A. I, III, IV

B. II and III

C. I and II

D. II, III, IV

21. Which does not affect enzyme rate? *(Average Rigor) (Skill 3.2)*

A. increase of temperature

B. amount of substrate

C. pH

D. size of the cell

22. Sulfur oxides and nitrogen oxides in the environment react with water to cause *(Rigorous) (Skill 3.15)*

A. ammonia

B. acidic precipitation

C. sulfuric acid

D. global warming

23. The loss of an electron is _____ and the gain of an electron is _____. *(Rigorous) (Skill 3.4)*

A. oxidation, reduction

B. reduction, oxidation

C. glycolysis, photosynthesis

D. photosynthesis, glycolysis

24. The product of anaerobic respiration in animals is *(Rigorous) (Skill 3.5)*

A. carbon dioxide

B. lactic acid

C. pyruvate

D. ethyl alcohol

25. **In the comparison of respiration to photosynthesis, which statement is true?**
(Rigorous)(Skill 3.6)

   A. oxygen is a waste product in photosynthesis but not in respiration

   B. glucose is produced in respiration but not in photosynthesis

   C. carbon dioxide is formed in photosynthesis but not in respiration

   D. water is formed in respiration and also in photosynthesis

26. **Carbon dioxide is fixed in the form of glucose in**
(Rigorous) (Skill 3.6)

   A. Krebs cycle

   B. the light reactions

   C. the dark reactions (Calvin cycle)

   D. glycolysis

27. **During the Krebs cycle, 8 carrier molecules are formed. What are they?**
(Rigorous) (Skill 3.4)

   A. 3 NADH, 3 FADH, 2 ATP

   B. 6 NADH and 2 ATP

   C. 4 $FADH_2$ and 4 ATP

   D. 6 NADH and 2 $FADH_2$

28. **Which of the following is not post-transcriptional processing?**
(Rigorous) (Skill 5.3)

   A. 5' capping

   B. intron splicing

   C. polypeptide splicing

   D. 3' polyadenylation

29. **Polymerase chain reaction**
(Rigorous) (Skill 5.6)

   A. is a group of polymerases

   B. technique for amplifying DNA

   C. primer for DNA synthesis

   D. synthesis of polymerase

30. **Homozygous individuals**
    (Average Rigor) (Skill 5.9)

    A. have two different alleles

    B. are of the same species

    C. have the same features

    D. have a pair of identical alleles

31. **The two major ways to determine taxonomic classification are**
    (Average Rigor) (Skill 10.3)

    A. evolution and phylogeny

    B. reproductive success and evolution

    C. phylogeny and morphology

    D. size and color

32. **Man's scientific name is Homo sapiens. Choose the proper classification beginning with kingdom and ending with order.**
    (Rigorous) (Skill 10.4)

    A. Animalia, Vertebrata, Mammalia, Primate, Hominidae

    B. Animalia, Vertebrata, Chordata, Mammalia, Primate

    C. Animalia, Chordata, Vertebrata, Mammalia, Primate

    D. Chordata, Vertebrata, Primate, Homo, sapiens

33. **The scientific name Canis familiaris refers to the animal's**
    (Average Rigor) (Skill 10.4)

    A. kingdom and phylum names

    B. genus and species names

    C. class and species names

    D. order and family names

34. **Members of the same species** *(Average Rigor) (Skill 10.4)*

    A. look identical

    B. never change

    C. reproduce successfully among their group

    D. live in the same geographic location

35. **What is necessary for diffusion to occur?** *(Rigorous) (Skill 3.8)*

    A. carrier proteins

    B. energy

    C. a concentration gradient

    D. a membrane

36. **Which is an example of the use of energy to move a substance through a membrane from areas of low concentration to areas of high concentration?** *(Rigorous) (Skill 4.8)*

    A. osmosis

    B. active transport

    C. exocytosis

    D. phagocytosis

37. **A plant cell is placed in salt water. The resulting movement of water out of the cell is called** *(Rigorous) (Skill 4.8)*

    A. facilitated diffusion

    B. diffusion

    C. transpiration

    D. osmosis

38. **As the amount of waste production increases in a cell, the rate of excretion** *(Rigorous) (Skill 3.12)*

    A. slowly decreases

    B. remains the same

    C. increases

    D. stops due to cell death

39. **A type of molecule not found in the membrane of an animal cell is** *(Rigorous) (Skill 4.6)*

    A. phospholipid

    B. protein

    C. cellulose

    D. cholesterol

40. **Which type of cell would contain the most mitochondria?** *(Rigorous)(Skill 4.3)*

A. muscle cell

B. nerve cell

C. epithelium

D. blood cell

41. **The first cells that evolved on earth were probably of which type?** *(Rigorous) (Skill 4.2)*

A. autotrophs

B. eukaryotes

C. heterotrophs

D. prokaryotes

42. **According to the fluid-mosaic model of the cell membrane, membranes are composed of** *(Average Rigor) (Skill 4.7)*

A. phospholipid bilayers with proteins embedded in the layers

B. one layer of phospholipids with cholesterol embedded in the layer

C. two layers of protein with lipids embedded the layers

D. DNA and fluid proteins

43. **All the following statements regarding both a mitochondria and a chloroplast are correct except** *(Rigorous) (Skill 4.5)*

A. they both transfer energy over a gradient

B. they both have DNA and are capable of reproduction

C. they both transfer light energy to chemical energy

D. they both make ATP

44. **This stage of mitosis includes cytokinesis or division of the cytoplasm and its organelles** *(Average Rigor) (Skill 4.4)*

A. anaphase

B. interphase

C. prophase

D. telophase

45. **Replication of chromosomes occurs during which phase of the cell cycle?** *(Rigorous) (Skill 4.4)*

A. prophase

B. interphase

C. metaphase

D. anaphase

46. **Which statement regarding mitosis is correct?**
(Rigorous) (Skill 4.4)

   A. diploid cells produce haploid cells for sexual reproduction

   B. sperm and egg cells are produced

   C. diploid cells produce diploid cells for growth and repair

   D. it allows for greater genetic diversity

47. **In a plant cell, telophase is described as**
(Average Rigor) (Skill 4.4)

   A. the time of chromosome doubling

   B. cell plate formation

   C. the time when crossing over occurs

   D. cleavage furrow formation

48. **Identify this stage of mitosis**
(Average Rigor) (Skill 4.4)

   A. anaphase

   B. metaphase

   C. telophase

   D. prophase

49. **Identify this stage of mitosis**
(Average Rigor) (Skill 4.4)

   A. prophase

   B. telophase

   C. anaphase

   D. metaphase

50. **Identify this stage of mitosis**
(Average Rigor) (Skill 4.4)

A. anaphase

B. metaphase

C. prophase

D. telophase

51. **Oxygen is given off in the**
(Rigorous) (Skill 3.6)

A. light reactions of
photosynthesis

B. dark reactions of
photosynthesis

C. Kreb's cycle

D. reduction of NAD+ to NADH

52. **In the electron transport chain,
all the following are true except**
(Rigorous) (Skill 3.4)

A. it occurs in the mitochondrion

B. it does not make ATP directly

C. the net gain of energy is 30
ATP

D. most molecules in the
electron transport chain are
proteins.

53. **The area of a DNA nucleotide
that varies is the**
(Average Rigor) (Skill 3.4)

A. deoxyribose

B. phosphate group

C. nitrogen base

D. sugar

54. **A DNA strand has the base
sequence of TCAGTA. Its DNA
complement would have the
following sequence**
(Average Rigor) (Skill 5.3)

A. ATGACT

B. TCAGTA

C. AGUCAU

D. AGTCAT

55. **Genes function in specifying
the structure of which
molecule?**
(Average Rigor) (Skill 5.4)

A. carbohydrates

B. lipids

C. nucleic acids

D. proteins

56. **What is the correct order of steps in protein synthesis?** *(Rigorous) (Skill 5.3)*

    A. transcription, then replication

    B. transcription, then translation

    C. translation, then transcription

    D. replication, then translation

57. **This carries amino acids to the ribosome in protein synthesis** *(Average Rigor) (Skill 5.3)*

    A. messenger RNA

    B. ribosomal RNA

    C. transfer RNA

    D. DNA

58. **A protein is sixty amino acids in length. This requires a coded DNA sequence of how many nucleotides?** *(Rigorous) (Skill 5.3)*

    A. 20

    B. 30

    C. 120

    D. 180

59. **A DNA molecule has the sequence of ACTATG. What is the anticodon of this molecule?** *(Rigorous) (Skill 5.3)*

    A. UGAUAC

    B. ACUAUG

    C. TGATAC

    D. ACTATG

60. **The term "phenotype" refers to which of the following?** *(Average Rigor) (Skill 5.9)*

    A. a condition which is heterozygous

    B. the genetic makeup of an individual

    C. a condition which is homozygous

    D. how the genotype is expressed

61. **The ratio of brown-eyed to blue-eyed children from the mating of a blue-eyed male to a heterozygous brown-eyed female would be expected to be which of the following?** *(Rigorous) (Skill 5.9)*

    A. 2:1

    B. 1:1

    C. 1:0

    D. 1:2

62. **The Law of Segregation defined by Mendel states that**
(Average Rigor) (Skill 5.9)

A. when sex cells form, the two alleles that determine a trait will end up on different gametes

B. only one of two alleles is expressed in a heterozygous organism

C. the allele expressed is the dominant allele

D. alleles of one trait do not affect the inheritance of alleles on another chromosome

63. **When a white flower is crossed with a red flower, incomplete dominance can be seen by the production of which of the following?**
(Average Rigor) (Skill 5.9)

A. pink flowers

B. red flowers

C. white flowers

D. red and white flowers

64. **Sutton observed that genes and chromosomes behaved the same. This led him to his theory which stated**
(Rigorous) (Skill 5.12)

A. that meiosis causes chromosome separation

B. that linked genes are able to separate

C. that genes and chromosomes have the same function

D. that genes are found on chromosomes

65. **Amniocentesis is**
(Rigorous) (Skill 5.11)

A. a non-invasive technique for detecting genetic disorders

B. a bacterial infection

C. extraction of amniotic fluid

D. removal of fetal tissue

66. A child with type O blood has a father with type A blood and a mother with type B blood. The genotypes of the parents respectively would be which of the following?
(Rigorous) (Skill 10.5)

   A. AA and BO

   B. AO and BO

   C. AA and BB

   D. AO and OO

67. Any change that affects the sequence of bases in a gene is called a (n)
(Rigorous) (Skill 10.6)

   A. deletion

   B. polyploid

   C. mutation

   D. duplication

68. The *lac* operon

   I. contains the *lac Z, lac Y, lac A* genes
   II. converts glucose to lactose
   III. contains a repressor
   IV. is on when the repressor is Activated
(Rigorous) (Skill 5.5)

   A. I

   B. II

   C. III and IV

   D. I and III

69. Which of the following factors will affect the Hardy-Weinberg law of equilibrium, leading to evolutionary change?
(Rigorous) (Skill 10.12)

   A. no mutations

   B. non-random mating

   C. no immigration or emigration

   D. Large population

70. If a population is in Hardy-Weinberg equilibrium and the frequency of the recessive allele is .3, what percentage of the population would be expected to be heterozygous? *(Rigorous) (Skill 10.12)*

    A. 9%

    B. 49%

    C. 42%

    D. 21%

71. Crossing over, which increases genetic diversity occurs during which stage(s)? *(Rigorous) (Skill 5.8)*

    A. telophase II in meiosis

    B. metaphase in mitosis

    C. interphase in both mitosis and meiosis

    D. prophase I in meiosis

72. Cancer cells divide extensively and invade other tissues. This continuous cell division is due to *(Rigorous) (Skill 5.7)*

    A. density dependent inhibition

    B. lack of density dependent inhibition

    C. chromosome replication

    D. Growth factors

73. Which process(es) results in a haploid chromosome number? *(Rigorous) (Skill 5.8)*

    A. both meiosis and mitosis

    B. mitosis

    C. meiosis

    D. replication and division

74. Segments of DNA can be transferred from the DNA of one organism to another through the use of which of the following? *(Average Rigor) (Skill 10.6)*

    A. bacterial plasmids

    B. tRNA molecules

    C. chromosomes from frogs

    D. plant DNA

75. Which of the following is not true regarding restriction enzymes? *(Rigorous) (Skill 5.6)*

    A. they do not aid in recombination procedures

    B. they are used in genetic engineering

    C. they are named after the bacteria in which they naturally occur

    D. they identify and splice certain base sequences on DNA

76. A virus that can remain dormant until a certain environmental condition causes its rapid increase is said to be
(Average Rigor) (Skill 6.1)

    A. lytic

    B. benign

    C. saprophytic

    D. lysogenic

77. Which is not considered to be a morphological type of bacteria?
(Average Rigor) (Skill 6.3)

    A. obligate

    B. coccus

    C. spirillum

    D. bacillus

78. Antibiotics are effective in fighting bacterial infections due to their ability to
(Rigorous) (Skill 6.5)

    A. interfere with DNA replication in the bacteria

    B. prevent the formation of new cell walls in the bacteria

    C. disrupt the ribosome of the bacteria

    D. All of the above.

79. Bacteria commonly reproduce by a process called binary fission. Which of the following best defines this process?
(Average Rigor) (Skill 4.3)

    A. viral vectors carry DNA to new bacteria

    B. DNA from one bacterium enters another

    C. DNA doubles and the bacterial cell divides

    D. DNA from dead cells is absorbed into bacteria

80. All of the following are examples of a member of Kingdom Fungi except
(Average Rigor) (Skill 4.6)

    A. mold

    B. algae

    C. mildew

    D. mushrooms

81. Protists are classified into major groups according to
(Average Rigor) (Skill 4.6)

    A. their method of obtaining nutrition

    B. reproduction

    C. metabolism

    D. their form and function

82. In comparison to protist cells, moneran cells

    I.   are usually smaller
    II. evolved later
    III. are more complex
    IV. contain more organelles
    (Rigorous) (Skill 6.1)

    A. I

    B. I and II

    C. II and III

    D. I and IV

83. Spores characterize the reproduction mode for which of the following group of plants?
    (Average Rigor) (Skill 7.9)

    A. algae

    B. flowering plants

    C. conifers

    D. ferns

84. Water movement to the top of a twenty foot tree is most likely due to which principle?
    (Rigorous) (Skill 3.1)

    A. osmotic pressure

    B. xylem pressure

    C. capillarity

    D. transpiration

85. What are the stages of development from the egg to the plant?
    (Average Rigor) (Skill 7.9)

    A. morphogenesis, growth, and cellular differentiation

    B. cell differentiation, growth, and morphogenesis

    C. growth, morphogenesis, and cellular differentiation

    D. growth, cellular differentiation, and morphogenesis

86. In angiosperms, the food for the developing plant is found in which of the following structures?
    (Average Rigor) (Skill 7.10)

    A. ovule

    B. endosperm

    C. male gametophyte

    D. cotyledon

87. The process in which pollen grains are released from the anthers is called
    (Easy) (Skill 7.9)

    A. pollination

    B. fertilization

    C. blooming

    D. dispersal

88. **Which of the following is not a characteristic of a monocot?** *(Average Rigor) (Skill 7.6)*

A. parallel veins in leaves

B. petals of flowers occur in multiples of 4 or 5

C. one seed leaf

D. vascular tissue scattered throughout the stem

89. **What controls gas exchange on the bottom of a plant leaf?** *(Easy) (Skill 7.3)*

A. stomata

B. epidermis

C. collenchyma and schlerenchyma

D. palisade mesophyll

90. **How are angiosperms different from other groups of plants?** *(Average Rigor) (Skill 7.5)*

A. presence of flowers and fruits

B. production of spores for reproduction

C. true roots and stems

D. seed production

91. **Generations of plants alternate between** *(Average Rigor) (Skill 7.7)*

A. angiosperms and bryophytes

B. flowering and nonflowering stages

C. seed bearing and spore bearing plants

D. haploid and diploid stages

92. **Double fertilization refers to which choice of the following?** *(Average Rigor) (Skill 7.9)*

A. two sperm fertilizing one egg

B. fertilization of a plant by gametes from two separate plants

C. two sperm enter the plant embryo sac; one sperm fertilizes the egg, the other forms the endosperm

D. the production of non-identical twins through fertilization of two separate eggs

93. **Characteristics of coelomates include:**

    I. no true digestive system
    II. two germ layers
    III. true fluid filled cavity
    IV. three germ layers
    (Rigorous) (Skill 8.2)

    A. I

    B. II and IV

    C. IV

    D. III and IV

94. **Which phylum accounts for 85% of all animal species?**
    (Rigorous) (Skill 8.2)

    A. Nematoda

    B. Chordata

    C. Arthropoda

    D. Cnidaria

95. **Which is the correct statement regarding the human nervous system and the human endocrine system?**
    (Rigorous) (Skill 8.6)

    A. the nervous system maintains homeostasis whereas the endocrine system does not

    B. endocrine glands produce neurotransmitters whereas nerves produce hormones

    C. nerve signals travel on neurons whereas hormones travel through the blood

    D. the nervous system involves chemical transmission whereas the endocrine system does not

96. **A muscular adaptation to move food through the digestive system is called**
    (Average Rigor) (Skill 8.10)

    A. peristalsis

    B. passive transport

    C. voluntary action

    D. bulk transport

97. The role of neurotransmitters in nerve action is
(Rigorous) (Skill 8.6)

A. turn off sodium pump

B. turn off calcium pump

C. send impulse to neuron

D. send impulse to the body

98. Fats are broken down by which substance?
(Rigorous) (Skill 3.4)

A. bile produced in the gall bladder

B. lipase produced in the gall bladder

C. glucagons produced in the liver

D. bile produced in the liver

99. Fertilization in humans usually occurs in the
(Average Rigor) (Skill 4.4)

A. uterus

B. ovary

C. fallopian tubes

D. vagina

100. All of the following are found in the dermis layer of skin except
(Average Rigor) (Skill 8.7)

A. sweat glands

B. keratin

C. hair follicles

D. blood vessels

101. Which is the correct sequence of embryonic development in a frog?
(Rigorous) (Skill 1.17)

A. cleavage – blastula – gastrula

B. cleavage – gastrula – blastula

C. blastula – cleavage – gastrula

D. gastrula – blastula – cleavage

102. Food is carried through the digestive tract by a series of wave-like contractions. This process is called
(Average Rigor) (Skill 8.5)

A. peristalsis

B. chyme

C. digestion

D. absorption

103. Movement is possible by the action of muscles pulling on *(Easy) (Skill 8.7)*

A. skin

B. bones

C. joints

D. ligaments

104. All of the following are functions of the skin except *(Easy) (Skill 8.9)*

A. storage

B. protection

C. sensation

D. regulation of temperature

105. Hormones are essential to the regulation of reproduction. What organ is responsible for the release of hormones for sexual maturity? *(Rigorous) (Skill 8.10)*

A. pituitary gland

B. hypothalamus

C. pancreas

D. thyroid gland

106. A bicyclist has a heart rate of 110 beats per minute and a stroke volume of 85 mL per beat. What is the cardiac output? *(Rigorous) (Skill 8.4)*

A. 9.35 L/min

B. 1.29 L/min

C. 0.772 L/min

D. 129 L/min

107. After sea turtles are hatched on the beach, they start the journey to the ocean. This is due to *(Easy) (Skill 8.12)*

A. innate behavior

B. territoriality

C. the tide

D. learned behavior

108. A school age boy had the chicken pox as a baby. He will most likely not get this disease again because of *(Easy) (Skill 3.10)*

A. passive immunity

B. vaccination

C. antibiotics

D. active immunity

109. High humidity and temperature stability are present in which of the following biomes?
(Average Rigor) (Skill 9.1)

   A. taiga

   B. deciduous forest

   C. desert

   D. tropical rain forest

110. The biological species concept applies to
(Average Rigor) (Skill 10.2)

   A. asexual organisms

   B. extinct organisms

   C. sexual organisms

   D. fossil organisms

111. Which term is not associated with the water cycle?
(Average Rigor) (Skill 9.3)

   A. precipitation

   B. transpiration

   C. fixation

   D. evaporation

112. All of the following are density independent factors that affect a population except
(Average Rigor)(Skill 9.5)

   A. temperature

   B. rainfall

   C. predation

   D. soil nutrients

113. In the growth of a population, the increase is exponential until carrying capacity is reached. This is represented by a (n)
(Average Rigor) (Skill 9.5)

   A. S curve

   B. J curve

   C. M curve

   D. L curve

114. Primary succession occurs after
(Average Rigor) (Skill 9.7)

   A. nutrient enrichment

   B. a forest fire

   C. bare rock is exposed after a water table recedes

   D. a housing development is built

115. Crabgrass – grasshopper – frog – snake – eagle. If DDT were present in an ecosystem, which organism would have the highest concentration in its system?
(Rigorous) (Skill 9.13)

A. grasshopper

B. eagle

C. frog

D. crabgrass

116. Which trophic level has the highest ecological efficiency?
(Average Rigor) (Skill 9.4)

A. decomposers

B. producers

C. tertiary consumers

D. secondary consumers

117. A clownfish is protected by the sea anemone's tentacles. In turn, the anemone receives uneaten food from the clownfish. This is an example of
(Easy) (Skill 9.6)

A. mutualism

B. parasitism

C. commensalisms

D. competition

118. If the niches of two species overlap, what usually results?
(Average Rigor) (Skill 7.4)

A. a symbiotic relationship

B. cooperation

C. competition

D. a new species

119. Oxygen created in photosynthesis comes from the breakdown of
(Rigorous) (Skill 3.6)

A. carbon dioxide

B. water

C. glucose

D. carbon monoxide

120. Which photosystem makes ATP?
(Average Rigor) (Skill 3.6)

A. photosystem I

B. photosystem II

C. photosystem III

D. photosystem IV

121. All of the following gasses made up the primitive atmosphere except
(Average Rigor)(Skill 3.13)

   A. ammonia

   B. methane

   C. oxygen

   D. hydrogen

122. The Endosymbiotic theory states that
(Average Rigor) (Skill 4.3)

   A. eukaryotes arose from prokaryotes

   B. animals evolved in close relationships with one another

   C. the prokaryotes arose from eukaryotes

   D. life arose from inorganic compounds

123. Which aspect of science does not support evolution?
(Easy) (Skill 1.17)

   A. comparative anatomy

   B. organic chemistry

   C. comparison of DNA among organisms

   D. analogous structures

124. Evolution occurs in
(Easy) (Skill 1.17)

   A. individuals

   B. populations

   C. organ systems

   D. cells

125. Which process contributes to the large variety of living things in the world today?
(Average Rigor) (Skill 1.17)

   A. meiosis

   B. asexual reproduction

   C. mitosis

   D. alternation of generations

126. The wing of bird, human arm and whale flipper have the same bone structure. These are called
(Easy) (Skill 1.17)

   A. polymorphic structures

   B. homologous structures

   C. vestigial structures

   D. analogous structures

127. Which biome is the most prevalent on Earth?
(Easy) (Skill 9.1)

A. marine

B. desert

C. savanna

D. tundra

128. Which of the following is not an abiotic factor?
(Average Rigor) (Skill 9.7)

A. temperature

B. rainfall

C. soil quality

D. bacteria

129. DNA synthesis results in a strand that is synthesized continuously. This is the
(Rigorous) (Skill 5.2)

A. lagging strand

B. leading strand

C. template strand

D. complementary strand

130. Using a gram staining technique, it is observed that E. coli stains pink. It is therefore
(Easy) (Skill 6.1)

A. gram positive

B. dead

C. gram negative

D. gram neutral

131. A light microscope has an ocular of 10X and an objective of 40X. What is the total magnification?
(Easy) (Skill 1.1)

A. 400X

B. 30X

C. 50X

D. 4000X

132. Three plants were grown. The following data was taken. Determine the mean growth.
Plant 1: 10cm Plant 2: 20cm
Plant 3: 15cm
(Rigorous) (Skill 1.4)

A. 5 cm

B. 45 cm

C. 12 cm

D. 15 cm

133. Electrophoresis separates DNA on the basis of
(Average Rigor) (Skill 1.5)

A. amount of current

B. molecular size

C. positive charge of the molecule

D. solubility of the gel

134. The reading of a meniscus in a graduated cylinder is done at the
(Easy) (Skill 1.3)

A. top of the meniscus

B. middle of the meniscus

C. bottom of the meniscus

D. closest whole number

135. Two hundred plants were grown. Fifty plants died. What percentage of the plants survived?
(Easy) (Skill 1.4)

A. 40%

B. 25%

C. 75%

D. 50%

136. Which is not a correct statement regarding the use of a light microscope?
(Easy) (Skill 1.1)

A. carry the microscope with two hands

B. store on the low power objective

C. clean all lenses with lens paper

D. Focus first on high power

137. Spectrophotometry utilizes the principle of
(Easy) (Skill 1.5)

A. light transmission

B. molecular weight

C. solubility of the substance

D. electrical charges

138. Paper chromotography is most often associated with the separation of
(Average Rigor) (Skill 1.5)

A. nutritional elements

B. DNA

C. proteins

D. plant pigments

139. A genetic engineering advancement in the medical field is
(Easy) (Skill 5.6)

   A. gene therapy

   B. pesticides

   C. degradation of harmful chemicals

   D. antibiotics

140. Which scientists are credited with the discovery of the structure of DNA?
(Easy) (Skill 1.14)

   A. Hershey & Chase

   B. Sutton & Morgan

   C. Watson & Crick

   D. Miller & Fox

141. Negatively charged particles that circle the nucleus of an atom are called
(Easy) (Skill 1.3)

   A. neutrons

   B. neutrinos

   C. electrons

   D. protons

142. The shape of a cell depends on its
(Easy) (Skill 4.2)

   A. function

   B. structure

   C. age

   D. size

143. The most ATP is generated through
(Average Rigor)(Skill 3.4)

   A. fermentation

   B. glycolysis

   C. chemiosmosis

   D. Krebs cycle

144. In DNA, adenine bonds with _____, while cytosine bonds with _____.
(Average Rigor) (Skill 5.2)

   A. thymine/guanine

   B. adenine/cytosine

   C. cytosine/adenine

   D. guanine/thymine

145. The individual parts of cells are best studied using a (n)
(Easy) (Skill 1.2)

A. ultracentrifuge

B. phase-contrast microscope

C. CAT scan

D. electron microscope

146. Thermoacidophiles are
(Rigorous) (Skill 6.1)

A. prokaryotes

B. eukaryotes

C. bacteria

D. archaea

147. Which of the following is not a type of fiber that makes up the cytoskeleton?
(Rigorous) (Skill 4.2)

A. vacuoles

B. microfilaments

C. microtubules

D. intermediate filaments

148. Viruses are made of
(Easy) (Skill 5.7)

A. a protein coat surrounding a nucleic acid

B. DNA, RNA and a cell wall

C. a nucleic acid surrounding a protein coat

D. protein surrounded by DNA

149. Reproductive isolation results in
(Easy) (Skill 10.2)

A. extinction

B. migration

C. follilization

D. speciation

150. This protein structure consists of the coils and folds of polypeptide chains. Which is it?
(Average Rigor) (Skill 3.1)

A. secondary structure

B. quaternary structure

C. tertiary structure

D. primary structure

ANSWER KEY

1.	C	31.	C	61.	B	91.	D	121.	C
2.	C	32.	C	62.	A	92.	C	122.	A
3.	B	33.	B	63.	A	93.	D	123.	B
4.	D	34.	C	64.	D	94.	C	124.	B
5.	C	35.	C	65.	C	95.	C	125.	A
6.	D	36.	B	66.	B	96.	A	126.	B
7.	D	37.	D	67.	C	97.	A	127.	A
8.	C	38.	C	68.	D	98.	D	128.	D
9.	D	39.	C	69.	B	99.	C	129.	B
10.	D	40.	A	70.	C	100.	B	130.	C
11.	A	41.	D	71.	D	101.	A	131.	A
12.	A	42.	A	72.	B	102.	A	132.	D
13.	B	43.	C	73.	C	103.	B	133.	B
14.	C	44.	D	74.	A	104.	A	134.	C
15.	C	45.	B	75.	A	105.	B	135.	C
16.	A	46.	C	76.	D	106.	A	136.	D
17.	A	47.	B	77.	A	107.	A	137.	A
18.	C	48.	B	78.	D	108.	D	138.	D
19.	D	49.	B	79.	C	109.	D	139.	A
20.	A	50.	A	80.	B	110.	C	140.	C
21.	D	51.	A	81.	A	111.	C	141.	C
22.	B	52.	C	82.	A	112.	C	142.	A
23.	A	53.	C	83.	D	113.	A	143.	C
24.	B	54.	D	84.	D	114.	C	144.	A
25.	A	55.	D	85.	C	115.	B	145.	D
26.	C	56.	B	86.	B	116.	B	146.	D
27.	D	57.	C	87.	A	117.	A	147.	A
28.	C	58.	D	88.	B	118.	C	148.	A
29.	B	59.	B	89.	A	119.	B	149.	D
30.	D	60.	D	90.	A	120.	A	150.	A

Rigor Table

	Easy %20	Average Rigor %40	Rigorous %40
Question #	6,7, 8, 10, 11, 14, 87, 89, 103, 104, 107, 108, 117, 123, 124, 126, 127, 130, 131, 134, 135, 136, 137, 139, 140, 141, 142, 145, 148, 149	1, 4, 12, 13, 15, 16, 17, 21, 30, 31, 33, 34, 42, 44, 47, 48, 49, 50, 53, 54, 55, 57, 60, 62, 63, 74, 76, 77, 79, 80, 81, 83, 85, 86, 88, 90, 91, 92, 96, 99, 100, 102, 109, 110, 111, 112, 113, 114, 116, 118, 120, 121, 122, 125, 128, 133, 138, 143, 144, 150	2, 3, 5, 9, 18, 19, 20, 22, 24, 25, 26, 27, 28, 29, 32, 35, 36, 37, 38, 39, 40, 41, 43, 45, 46, 51, 52, 56, 58, 59, 61, 64, 65, 66, 67, 68, 69, 70, 71, 72, 73, 75, 78, 82, 84, 93, 94, 95, 97, 98, 101, 105, 106, 115, 119, 129, 132, 146, 147

RATIONALE

1. **A student designed a science project testing the effects of light and water on plant growth. You would recommend that she**
   (Average Rigor) (Skill 1.12)

   A.   manipulate the temperature as well.

   B.   also alter the pH of the water as another variable.

   C.   omit either water or light as a variable.

   D.   also alter the light concentration as another variable.

Answer: C. omit either water or light as a variable
In science, experiments should be designed so that only one variable is manipulated at a time.

2. **Identify the control in the following experiment. A student had four plants grown under the following conditions and was measuring photosynthetic rate by measuring mass. 2 plants in 50% light and 2 plants in 100% light.**
   (Rigorous) (Skill 1.16)

   A.   plants grown with no added nutrients

   B.   plants grown in the dark

   C   plants in 100% light

   D.   plants in 50% light

Answer: C. plants in 100% light
The 100% light plants are those that the student will be comparing the 50% plants to. This will be the control.

3.   **In an experiment measuring the growth of bacteria at different temperatures, identify the independent variable.** *(Rigorous) (Skill 1.9)*

A.   growth of number of colonies

B.   temperature

C.   type of bacteria used

D.   light intensity

Answer: B. temperature
The independent variable is controlled by the experimenter. Here, the temperature is controlled to determine its effect on the growth of bacteria (dependent variable).

4.   **A scientific theory**
      (Average Rigor)(Skill 1.8)

A.   proves scientific accuracy.

B.   is never rejected.

C.   results in a medical breakthrough.

D.   may be altered at a later time.

Answer: D. may be altered at a later time
Scientific theory is usually accepted and verified information but can always be changed at anytime.

5. **Which is the correct order of methodology?**

   1) **testing revised explanation,**
   2) **setting up a controlled experiment to test explanation,**
   3) **drawing a conclusion,**
   4) **suggesting an explanation for observations, and**
   5) **compare observed results to hypothesized results**

   (Rigorous)(Skill 1.8)

   A.   4, 2, 3, 1, 5

   B.   3, 1, 4, 2, 5

   C.   4, 2, 5, 1, 3

   D.   2, 5, 4, 1, 3

Answer: C. 4, 2, 5, 1, 3
The first step in scientific inquiry is posing a question to be answered. Next, a hypothesis is formed to provide a plausible explanation. An experiment is then proposed and performed to test this hypothesis. A comparison between the predicted and observed results is the next step. Conclusions are then formed and it is determined whether the hypothesis is correct or incorrect. If incorrect, the next step is to form a new hypothesis and the process is repeated.

6. **Given a choice, which is the most desirable method of heating a substance in the lab?**
   (Easy) (Skill 2.6)

   A.   alcohol burner

   B.   gas burner

   C.   Bunsen burner

   D.   hot plate

Answer: D. hot plate
A hotplate is the only heat source from the choices above that does not have an open flame. The use of a hot plate will reduce the risk of fire and injury to students.

7.   **Biological waste should be disposed of**
     (Easy) (Skill 2.6)

     A.   in the trash can.

     B.   under a fume hood.

     C.   in the broken glass box.

     D.   in an autoclavable biohazard bag.

Answer: D. in an autoclavable biohazard bag
Biological material should never be stored near food or water used for human consumption. All biological material should be appropriately labeled. All blood and body fluids should be put in a well-contained container with a secure lid to prevent leaking. All biological waste should be disposed of in biological hazardous waste bags.

8.   **Chemicals should be stored**
     (Easy) (Skill 2.6)

     A.   in a cool dark room.

     B.   in a dark room.

     C.   according to their reactivity with other substances.

     D.   in a double locked room.

Answer: C. according to their reactivity with other substances
All chemicals should be stored with other chemicals of similar reactivity. Failure to do so could result in an undesirable chemical reaction.

9. **Given the choice of lab activities, which would you omit?**
   (Rigorous) (Skill 1.3)

   A.   a genetics experiment tracking the fur color of mice

   B.   dissecting a preserved fetal pig

   C.   a lab relating temperature to respiration rate using live goldfish.

   D.   pithing a frog to see the action of circulation

Answer: D. pithing a frog to see the action of circulation
The use of live vertebrate organisms in a way that may harm the animal is prohibited. The observation of fur color in mice is not harmful to the animal and the use of live goldfish is acceptable because they are invertebrates. The dissection of a fetal pig is acceptable if it comes from a known origin.

10. **Who should be notified in the case of a serious chemical spill?**

   I.   the custodian
   II. The fire department
   III. the chemistry teacher
   IV. the administration

   (Easy) (Skill 2.6)

   A.   I

   B.   II

   C.   II and III

   D.   II and IV

Answer: D. II and IV
For large spills, the school administration and the local fire department should be notified.

11.   The "Right to Know" law states
        (Easy) (Skill 2.6)

        A.   the inventory of toxic chemicals checked against the "Substance List" be available.

        B.   that students are to be informed on alternatives to dissection.

        C.   that science teachers are to be informed of student allergies.

        D.   that students are to be informed of infectious microorganisms used in lab.

Answer: A. the inventory of toxic chemicals checked against the "Substance List" be available
The right to know law pertains to chemical substances in the lab. Employees should check the material safety data sheets and the substance list for potential hazards in the lab.

12.   In which situation would a science teacher be liable?
        (Average Rigor) (Skill 2.6)

        A.   a teacher leaves to receive an emergency phone call and a student slips and falls.

        B.   a student removes their goggles and gets dissection fluid in their eye.

        C.   a faulty gas line results in a fire.

        D.   a students cuts themselves with a scalpel.

Answer: A. a teacher leaves to receive an emergency phone call and a student slips and falls
A teacher has an obligation to be present in the lab at all times. If the teacher needs to leave, an appropriate substitute is needed.

13.   **Which statement best defines negligence?**
       (Average Rigor) (Skill 2.6)

      A.   failure to give oral instructions for those with reading disabilities

      B.   failure to exercise ordinary care

      C.   inability to supervise a large group of students.

      D.   reasonable anticipation that an event may occur

Answer: B. failure to exercise ordinary care
Negligence is the failure to exercise ordinary or reasonable care.

14.   **Which item should always be used when using chemicals with noxious vapors?**
       (Easy) (Skill 2.6)

      A.   eye protection

      B.   face shield

      C.   fume hood

      D.   lab apron

Answer: C. fume hood
Fume hoods are designed to protect the experimenter from chemical fumes. The three other choices do not prevent chemical fumes from entering the respiratory system.

15.   Identify the correct sequence of organization of living things.
       (Average Rigor) (Skill 10.3)

       A.   cell – organelle – organ system – tissue – organ – organism

       B.   cell – tissue – organ – organ system – organelle – organism

       C.   organelle – cell – tissue – organ – organ system – organism

       D.   tissue – organelle – organ – cell – organism – organ system

Answer: C. organelle – cell – tissue – organ – organ system – organism
An organism, such as a human, is comprised of several organ systems such as
the circulatory and nervous systems. These organ systems consist of many
organs including the heart and the brain. These organs are made of tissue such
as cardiac muscle. Tissues are made up of cells, which contain organelles like
the mitochondria and the Golgi apparatus.

16.   Which is not a characteristic of living things?
       (Average Rigor) (Skill 4.2)

       A.   movement

       B.   cellular structure

       C.   metabolism

       D.   reproduction

Answer: A. movement
Movement is not a characteristic of life. Viruses are considered non-living
organisms but have the ability to move from cell to cell in its host organism.

17. **Which kingdom is comprised of organisms made of one cell with no nuclear membrane?**
   (Average Rigor) (Skill 4.2)

   A. Monera

   B. Protista

   C. Fungi

   D. Algae

Answer: A. monera
Monera is the only kingdom that is made up of unicellular organisms with no nucleus. Algae is a protist because it is made up of one type of tissue and it has a nucleus.

18. **Potassium chloride atoms are joined by a(n)**
   (Rigorous) (Skill 3.1)

   A. non polar covalent bond

   B. polar covalent bond

   C. ionic bond

   D. hydrogen bond

Answer: C. ionic bond
Ionic bonds are formed when one electron is stripped away from its atom to join another atom. Ionic compounds are called salts and potassium chloride is a salt; therefore, potassium chloride is an example of an ionic bond.

19. **Which of the following is a monomer?**
(Rigorous) (Skill 3.1)

    A.   RNA

    B.   glycogen

    C.   DNA

    D.   amino acid

Answer: D. amino acid
A monomer is the simplest unit of structure for a particular macromolecule. Amino acids are the basic unit that comprises a protein. RNA and DNA are polymers consisting of nucleotides and glycogen is a polymer consisting of many molecules of glucose.

20. **Which of the following are properties of water?**

    I. **High specific heat**
    II. **Strong ionic bonds**
    III. **Good solvent**
    IV. **High freezing point**
(Rigorous) (Skill 3.1)

    A.   I, III, IV

    B.   II and III

    C.   I and II

    D.   II, III, IV

Answer: A. I, III, IV
All are properties of water except strong ionic bonds. Water is held together by polar covalent bonds between hydrogen and oxygen.

21. **Which does not affect enzyme rate?**
    (*Average Rigor*) (*Skill 3.2*)

    A.   increase of temperature

    B.   amount of substrate

    C.   pH

    D.   size of the cell

Answer: D. size of the cell
Temperature and pH can affect the rate of reaction of an enzyme. The amount of substrate affects the enzyme as well. The enzyme acts on the substrate. The more substrate there is, the slower the enzyme rate. Therefore, the only chance left is D, the size of the cell, which has no effect on enzyme rate.

22. **Sulfur oxides and nitrogen oxides in the environment react with water to cause**
    (*Rigorous*) (*Skill 3.15*)

    A.   ammonia

    B.   acidic precipitation

    C.   sulfuric acid

    D.   global warming

Answer: B. acidic precipitation
Acidic precipitation is rain, snow, or fog with a pH less than 5.6. It is caused by sulfur oxides and nitrogen oxides that react with water in the air to form acids that fall down to Earth as precipitation.

23.   The loss of an electron is _____ and the gain of an electron is
      _____. *(Rigorous) (Skill 3.4)*

      A.   oxidation, reduction

      B.   reduction, oxidation

      C.   glycolysis, photosynthesis

      D.   photosynthesis, glycolysis

Answer: A. oxidation, reduction
Oxidation-reduction reactions are also known as redox reactions. In respiration, energy is released by the transfer of electrons by this process. The oxidation phase of this reaction is the loss of an electron and the reduction phase is the gain of an electron.

24.   The product of anaerobic respiration in animals is
      (Rigorous) (Skill 3.5)

      A.   carbon dioxide

      B.   lactic acid

      C.   pyruvate

      D.   ethyl alcohol

Answer: B. lactic acid
In anaerobic lactic acid fermentation, pyruvate is reduced by NADH to form lactic acid. This is the anaerobic process in animals. Alcoholic fermentation is the anaerobic process in yeast and some bacteria resulting in ethyl alcohol. Carbon dioxide and pyruvate are the products of aerobic respiration.

25.   **In the comparison of respiration to photosynthesis, which statement is true?** *(Rigorous) (Skill 3.6)*

   A.   oxygen is a waste product in photosynthesis but not in respiration

   B.   glucose is produced in respiration but not in photosynthesis

   C.   carbon dioxide is formed in photosynthesis but not in respiration

   D.   water is formed in respiration and also in photosynthesis

Answer: A. oxygen is a waste product in photosynthesis but not in respiration

In photosynthesis, water is split and the oxygen is given off as a waste product. In respiration, water and carbon dioxide are the waste products.

26.   **Carbon dioxide is fixed in the form of glucose in** *(Rigorous) (Skill 3.6)*

   A.   Krebs cycle

   B.   the light reactions

   C.   the dark reactions (Calvin cycle)

   D.   glycolysis

Answer: C. the dark reactions (Calvin cycle)

The ATP produced during the light reaction is needed to convert carbon dioxide to glucose in the Calvin cycle.

27.   During the Kreb's cycle, 8 carrier molecules are formed. What are they?
        (Rigorous) (Skill 3.4)

        A.   3 NADH, 3 FADH, 2 ATP

        B.   6 NADH and 2 ATP

        C.   4 FADH$_2$ and 4 ATP

        D.   6 NADH and 2 FADH$_2$

Answer: D. 6 NADH and 2 FADH$_2$
For each molecule of CoA that enters the Kreb's cycle, you get 3 NADH and 1 FADH$_2$. There are 2 molecules of CoA so the total yield is 6 NADH and 2 FADH$_2$ during the Krebs cycle.

28.   **Which of the following is not posttranscriptional processing?**
        (Rigorous) (Skill 5.3)

        A.   5' capping

        B.   intron splicing

        C.   polypeptide splicing

        D.   3' polyadenylation

Answer: C. polypeptide splicing
The removal of segments of polypeptides is a posttranslational process. The other three are methods of posttranscriptional processing.

29.   **Polymerase chain reaction** *(Rigorous) (Skill 5.6)*

        A. is a group of polymerases

        B. technique for amplifying DNA

        C. primer for DNA synthesis

        D. synthesis of polymerase

Answer: B. technique for amplifying DNA
PCR is a technique in which a piece of DNA can be amplified into billions of copies within a few hours.

30.   Homozygous individuals *(Average Rigor) (Skill 5.9)*

    A.   have two different alleles

    B.   are of the same species

    C.   have the same features

    D.   have a pair of identical alleles

Answer: D. have a pair of identical alleles
Homozygous individuals have a pair of identical alleles and heterozygous individuals have two different alleles.

31.   **The two major ways to determine taxonomic classification are *(Average Rigor) (Skill 10.3)***

    A.   evolution and phylogeny

    B.   reproductive success and evolution

    C.   phylogeny and morphology

    D.   size and color

Answer: C. phylogeny and morphology
Taxonomy is based on structure (morphology) and evolutionary relationships (phylogeny).

32.   **Man's scientific name is Homo sapiens. Choose the proper classification beginning with kingdom and ending with order. *(Rigorous) (Skill 10.4)***

    A.   Animalia, Vertebrata, Mammalia, Primate, Hominidae

    B.   Animalia, Vertebrata, Chordata, Mammalia, Primate

    C.   Animalia, Chordata, Vertebrata, Mammalia, Primate

    D.   Chordata, Vertebrata, Primate, Homo, sapiens

Answer: C. Animalia, Chordata, Vertebrata, Mammalia, Primate
The order of classification for humans is as follows: Kingdom, Animalia; Phylum, Chordata; Subphylum, Vertebrata; Class, Mammalia; Order, Primate; Family, Hominadae; Genus, Homo; Species, sapiens.

33. **The scientific name Canis familiaris refers to the animal's** *(Average Rigor) (Skill 10.4)*

    A.   kingdom and phylum names

    B.   genus and species names

    C.   class and species names

    D.   order and family names

Answer: B. genus and species names
Each species is scientifically known by a two-part name, or binomial. The first word in the name is the genus and the second word is its specific epithet (species name).

34. **Members of the same species** *(Average Rigor) (Skill 10.4)*

    A.   look identical

    B.   never change

    C.   reproduce successfully among their group

    D.   live in the same geographic location

Answer: C. reproduce successfully among their group
Species are defined by the ability to successfully reproduce with members of their own kind.

35. **What is necessary for diffusion to occur?** *(Rigorous) (Skill 3.8)*

    A.   carrier proteins

    B.   energy

    C.   a concentration gradient

    D.   a membrane

Answer: C. a concentration gradient
Diffusion is the ability of molecules to move from areas of high concentration to areas of low concentration (a concentration gradient).

36. Which is an example of the use of energy to move a substance through a membrane from areas of low concentration to areas of high concentration? *(Rigorous) (Skill 4.8)*

    A.   osmosis

    B.   active transport

    C.   exocytosis

    D.   phagocytosis

Answer: B. active transport
Active transport can move substances with or against the concentration gradient. This energy requiring process allows for molecules to move from areas of low concentration to high concentration areas.

37. A plant cell is placed in salt water. The resulting movement of water out of the cell is called *(Rigorous) (Skill 4.8)*

    A.   facilitated diffusion

    B.   diffusion

    C.   transpiration

    D.   osmosis

Answer: D. osmosis
Osmosis is simply the diffusion of water across a semi-permeable membrane. Water will diffuse out of the cell if there is less water on the outside of the cell.

38.   As the amount of waste production increases in a cell, the rate of excretion *(Rigorous) (Skill 3.12)*

A.   slowly decreases

B.   remains the same

C.   increases

D.   stops due to cell death

Answer: C. increases
Homeostasis is the control of the differences between internal and external environments. Excretion is the homeostatic system that regulates the amount of waste in a cell. As the amount of waste increases, the rate of excretion will increase to maintain homeostasis.

39.   A type of molecule not found in the membrane of an animal cell is *(Rigorous) (Skill 4.6)*

A.   phospholipid

B.   protein

C.   cellulose

D.   cholesterol

Answer: C. cellulose
Phospholipids, protein, and cholesterol are all found in animal cells. Cellulose, however, is only found in plant cells.

40.   Which type of cell would contain the most mitochondria? *(Rigorous) (Skill 4.3)*

A.   muscle cell

B.   nerve cell

C.   epithelium

D.   blood cell

Answer: A. muscle cell
Mitochondria are the site of cellular respiration where ATP is made. Muscle cells have the most mitochondria because they use a great deal of energy.

41.   The first cells that evolved on earth were probably of which type? *(Rigorous) (Skill 4.2)*

    A.   autotrophs

    B.   eukaryotes

    C.   heterotrophs

    D.   prokaryotes

Answer: D. prokaryotes
Prokaryotes date back to 3.5 billion years ago in the first fossil record. Their ability to adapt to the environment allows them to thrive in a wide variety of habitats.

42.   According to the fluid-mosaic model of the cell membrane, membranes are composed of *(Average Rigor) (Skill 4.7)*

    A.   phospholipid bilayers with proteins embedded in the layers

    B.   one layer of phospholipids with cholesterol embedded in the layer

    C.   two layers of protein with lipids embedded the layers

    D.   DNA and fluid proteins

Answer: A. phospholipid bilayers with proteins embedded in the layers
Cell membranes are composed of two phospholipids with their hydrophobic tails sandwiched between their hydrophilic heads, creating a lipid bilayer. The membrane contains proteins embedded in the layer (integral proteins) and proteins on the surface (peripheral proteins).

43.   All the following statements regarding both a mitochondria and a chloroplast are correct except *(Rigorous) (Skill 4.5)*

    A.   they both transfer energy over a gradient

    B.   they both have DNA and are capable of reproduction

    C.   they both transfer light energy to chemical energy

    D.   they both make ATP

Answer: C. they both transfer light energy to chemical energy

Cellular respiration does not transfer light energy to chemical energy. Cellular respiration transfers electrons to release energy. Photosynthesis utilizes light energy to produce chemical energy.

44.   **This stage of mitosis includes cytokinesis or division of the cytoplasm and its organelles** *(Average Rigor) (Skill 4.4)*

        A.   anaphase

        B.   interphase

        C.   prophase

        D.   telophase

Answer: D. telophase
The last stage of the mitotic phase is telophase. Here, the two nuclei form with a full set of DNA each. The cell is pinched into two cells and cytokinesis, or division of the cytoplasm and organelles, occurs.

45.   **Replication of chromosomes occurs during which phase of the cell cycle?** *(Rigorous) (Skill 4.4)*

        A.   prophase

        B.   interphase

        C.   metaphase

        D.   anaphase

Answer: B. interphase
Interphase is the stage where the cell grows and copies the chromosomes in preparation for the mitotic phase.

46.   Which statement regarding mitosis is correct? *(Rigorous) (Skill 4.4)*

A.   diploid cells produce haploid cells for sexual reproduction

B.   sperm and egg cells are produced

C.   diploid cells produce diploid cells for growth and repair

D.   it allows for greater genetic diversity

Answer: C. diploid cells produce diploid cells for growth and repair
The purpose of mitotic cell division is to provide growth and repair in body (somatic) cells. The cells begin as diploid and produce diploid cells.

47.   In a plant cell, telophase is described as *(Average Rigor) (Skill 4.4)*

A.   the time of chromosome doubling

B.   cell plate formation

C.   the time when crossing over occurs

D.   cleavage furrow formation

Answer: B. cell plate formation
During plant cell telophase, a cell plate is observed whereas a cleavage furrow is formed in animal cells.

48. **Identify this stage of mitosis** *(Average Rigor) (Skill 4.4)*

A. anaphase

B. metaphase

C. telophase

D. prophase

Answer: B. metaphase
During metaphase, the centromeres are at opposite ends of the cell. Here the chromosomes are aligned with one another.

49. **Identify this stage of mitosis** *(Average Rigor) (Skill 4.4)*

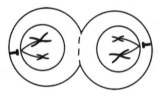

A. prophase

B. telophase

C. anaphase

D. metaphase

Answer: B. telophase
Telophase is the last stage of mitosis. Here, two nuclei become visible and the nuclear membrane resembles.

50.   Identify this stage of mitosis *(Average Rigor) (Skill 4.4)*

A.   anaphase

B.   metaphase

C.   prophase

D.   telophase

Answer: A. anaphase
During anaphase, the centromeres split in half and homologous chromosomes separate.

51.   Oxygen is given off in the *(Rigorous) (Skill 3.6)*

A.   light reactions of photosynthesis

B.   dark reactions of photosynthesis

C.   Krebs cycle

D.   reduction of NAD+ to NADH

Answer: A. light reactions of photosynthesis
The conversion of solar energy to chemical energy occurs in the light reactions. Electrons are transferred by the absorption of light by chlorophyll and causes water to split, releasing oxygen as a waste product.

52. **In the electron transport chain, all the following are true except** *(Rigorous) (Skill 3.4)*

    A.   it occurs in the mitochondrion

    B.   it does not make ATP directly

    C.   the net gain of energy is 30 ATP

    D.   most molecules in the electron transport chain are proteins.

Answer: C. the net gain of energy is 30 ATP
The end result of the electron transport chain is 34 molecules of ATP.

53. **The area of a DNA nucleotide that varies is the** *(Average Rigor) (Skill 3.4)*

    A.   deoxyribose

    B.   phosphate group

    C.   nitrogen base

    D.   sugar

Answer: C. nitrogen base
DNA is made of a 5 carbon sugar (deoxyribose), a phosphate group, and a nitrogenous base. There are four nitrogenous bases in DNA that allow for the four different nucleotides.

54. **A DNA strand has the base sequence of TCAGTA. Its DNA complement would have the following sequence** *(Average Rigor) (Skill 5.3)*

    A.   ATGACT

    B.   TCAGTA

    C.   AGUCAU

    D.   AGTCAT

Answer: D. AGTCAT
The complement strand to a single strand DNA molecule has a complementary sequence to the template strand. T pairs with A and C pairs with G. Therefore, the complement to TCAGTA is AGTCAT.

55. **Genes function in specifying the structure of which molecule? (Average Rigor) (Skill 5.4)**

    A.   carbohydrates

    B.   lipids

    C.   nucleic acids

    D.   proteins

Answer: D. proteins
Genes contain the sequence of nucleotides that code for amino acids. Amino acids are the building blocks of protein.

56. **What is the correct order of steps in protein synthesis? (Rigorous) (Skill 5.3)**

    A.   transcription, then replication

    B.   transcription, then translation

    C.   translation, then transcription

    D.   replication, then translation

Answer: B. transcription, then translation
A DNA strand first undergoes transcription to get a complementary mRNA strand. Translation of the mRNA strand then occurs to result in the tRNA adding the appropriate amino acid for an ending product of a protein.

57. **This carries amino acids to the ribosome in protein synthesis (Average Rigor) (Skill 5.3)**

    A.   messenger RNA

    B.   ribosomal RNA

    C.   transfer RNA

    D.   DNA

Answer: C. transfer RNA
The tRNA molecule is specific for a particular amino acid. The tRNA has an anticodon sequence that is complementary to the codon. This specifies where the tRNA places the amino acid in protein synthesis.

58.   A protein is sixty amino acids in length. This requires a coded DNA sequence of how many nucleotides? *(Rigorous) (Skill 5.3)*

    A.   20

    B.   30

    C.   120

    D.   180

Answer: D. 180
Each amino acid codon consists of 3 nucleotides. If there are 60 amino acids in a protein, then 60 x 30 = 180 nucleotides.

59.   A DNA molecule has the sequence of ACTATG. What is the anticodon of this molecule? *(Rigorous) (Skill 5.3)*

    A.   UGAUAC

    B.   ACUAUG

    C.   TGATAC

    D.   ACTATG

Answer: B. ACUAUG
The DNA is first transcribed into mRNA. Here, the DNA has the sequence ACTATG; therefore the complementary mRNA sequence is UGAUAC (remember, in RNA, T are U). This mRNA sequence is the codon. The anticodon is the complement to the codon. The anticodon sequence will be ACUAUG (remember, the anticodon is tRNA, so U is present instead of T).

60.   The term "phenotype" refers to which of the following? *(Average Rigor) (Skill 5.9)*

   A.   a condition which is heterozygous

   B.   the genetic makeup of an individual

   C.   a condition which is homozygous

   D.   how the genotype is expressed

Answer: D. how the genotype is expressed
Phenotype is the physical appearance of an organism due to its genetic makeup (genotype).

61.   The ratio of brown-eyed to blue-eyed children from the mating of a blue-eyed male to a heterozygous brown-eyed female would be expected to be which of the following? *(Rigorous) (Skill 5.9)*

   A.   2:1

   B.   1:1

   C.   1:0

   D.   1:2

Answer: B. 1:1
Use a Punnet square to determine the ratio.

	b	b
B	Bb	Bb
b	bb	bb

B = brown eyes, b = blue eyes

Female genotype is on the side and the male genotype is across the top.

The female is heterozygous and her phenotype is brown eyes. This means the dominant allele is for brown eyes. The male expresses the homozygous recessive allele for blue eyes. Their children are expected to have a ratio of brown eyes to blue eyes of 2:2; or 1:1.

62. **The Law of Segregation defined by Mendel states that** *(Average Rigor) (Skill 5.9)*

    A.   when sex cells form, the two alleles that determine a trait will end up on different gametes

    B.   only one of two alleles is expressed in a heterozygous organism

    C.   the allele expressed is the dominant allele

    D.   alleles of one trait do not affect the inheritance of alleles on another chromosome

Answer: A. when sex cells form, the two alleles that determine a trait will end up on different gametes
The law of segregation states that the two alleles for each trait segregate into different gametes.

63. **When a white flower is crossed with a red flower, incomplete dominance can be seen by the production of which of the following?** *(Average Rigor) (Skill 5.9)*

    A.   pink flowers

    B.   red flowers

    C.   white flowers

    D.   red and white flowers

Answer: A. pink flowers
Incomplete dominance is when the F_1 generation results in an appearance somewhere between the parents. Red flowers crossed with white flowers results in an F_1 generation with pink flowers.

64.   **Sutton observed that genes and chromosomes behaved the same. This led him to his theory which stated** *(Rigorous) (Skill 5.12)*

A.   that meiosis causes chromosome separation

B.   that linked genes are able to separate

C.   that genes and chromosomes have the same function

D.   that genes are found on chromosomes

Answer: D. that genes are found on chromosomes
Sutton observed how mitosis and meiosis confirmed Mendel's theory on "factors." His Chromosome Theory states that genes are located on chromosomes.

65.   **Amniocentesis is** *(Rigorous) (Skill 5.11)*

A.   a non-invasive technique for detecting genetic disorders

B.   a bacterial infection

C.   extraction of amniotic fluid

D.   removal of fetal tissue

Answer: C. extraction of amniotic fluid
Amniocentesis is a procedure in which a needle is inserted into the uterus to extract some of the amniotic fluid surrounding the fetus. Some genetic disorders can be detected by chemicals in the fluid.

66. **A child with type O blood has a father with type A blood and a mother with type B blood. The genotypes of the parents respectively would be which of the following?** *(Rigorous) (Skill 10.5)*

   A.   AA and BO

   B.   AO and BO

   C.   AA and BB

   D.   AO and OO

Answer: B. AO and BO
Type O blood has 2 recessive O genes. A child receives one allele from each parent; therefore each parent in this example must have an O allele. The father has type A blood with a genotype of AO and the mother has type B blood with a genotype of BO.

67. **Any change that affects the sequence of bases in a gene is called a(n)** *(Rigorous) (Skill 10.6)*

   A.   deletion

   B.   polyploid

   C.   mutation

   D.   duplication

Answer: C. mutation
A mutation is an inheritable change in DNA. They may be errors in replication or a spontaneous rearrangement of one or more segments of DNA. Deletion and duplication are type of mutations. Polyploidy is when an organism has more than two complete chromosome sets.

68.   The *lac* operon

      I. contains the *lac Z, lac Y, lac A* genes
      II. converts glucose to lactose
      III. contains a repressor
      IV. is on when the repressor is activated
      (Rigorous) (Skill 5.5)

      A.   I

      B.   II

      C.   III and IV

      D.   I and III

Answer: D. I and III
The *lac* operon contains the genes that encode for the enzymes used to convert lactose into fuel. It contains three genes: *lac A, lac Z,* and *lac Y*. It also contains a promoter and repressor. When the repressor is activated, the operon is off.

69.   Which of the following factors will affect the Hardy-Weinberg law of equilibrium, leading to evolutionary change? *(Rigorous) (Skill 10.12)*

      A.   no mutations

      B.   non-random mating

      C.   no immigration or emigration

      D.   Large population

Answer: B. non- random mating
There are five requirements to keep the Hardy-Weinberg equilibrium stable: no mutation, no selection pressures, an isolated population, a large population, and random mating.

70.   If a population is in Hardy-Weinberg equilibrium and the frequency of the recessive allele is 0.3, what percentage of the population would be expected to be heterozygous? *(Rigorous) (Skill 10.12)*

    A.   9%

    B.   49%

    C.   42%

    D.   21%

Answer: C. 42%

0.3 is the value of q. Therefore, $q^2 = 0.09$. According to the Hardy-Weinberg equation, $1 = p + q$.

$1 = p + 0.3$.
$p = 0.7$
$p^2 = 0.49$

Next, plug q^2 and p^2 into the equation $1 = p^2 + 2pq + q^2$.

$1 = 0.49 + 2pq + 0.09$ (where 2pq is the number of heterozygotes).
$1 = 0.58 + 2pq$
$2pq = 0.42$

Multiply by 100 to get the percent of heterozygotes to get 42%.

71. **Crossing over, which increases genetic diversity occurs during which stage(s)?** *(Rigorous) (Skill 5.8)*

    A.   telophase II in meiosis

    B.   metaphase in mitosis

    C.   interphase in both mitosis and meiosis

    D.   prophase I in meiosis

Answer: D. prophase I in meiosis
During prophase I of meiosis, the replicated chromosomes condense and pair with homologues in a process called synapsis. Crossing over, the exchange of genetic material between homologues to further increase diversity, occurs during prophase I.

72. **Cancer cells divide extensively and invade other tissues. This continuous cell division is due to** *(Rigorous) (Skill 5.7)*

    A.   density dependent inhibition

    B.   lack of density dependent inhibition

    C.   chromosome replication

    D.   Growth factors

Answer: B. lack of density dependent inhibition
Density dependent inhibition is when the cells crowd one another and consume all the nutrients; therefore halting cell division. Cancer cells, however, are density independent; meaning they can divide continuously as long as nutrients are present, and their growth is NOT inhibited by density

73. **Which process(es) results in a haploid chromosome number?**
    (Rigorous) (Skill 5.8)

    A.   both meiosis and mitosis

    B.   mitosis

    C.   meiosis

    D.   replication and division

Answer: C. meiosis
In meiosis, there are two consecutive cell divisions resulting in the reduction of the chromosome number by half (diploid to haploid).

74. **Segments of DNA can be transferred from the DNA of one organism to another through the use of which of the following?** *(Average Rigor) (Skill 10.6)*

    A.   bacterial plasmids

    B.   tRNA molecules

    C.   chromosomes from frogs

    D.   plant DNA

Answer: A. bacterial plasmids
Plasmids can transfer themselves (and therefore their genetic information) by a process called conjugation. This requires cell-cell contact.

75. **Which of the following is not true regarding restriction enzymes?**
    (Rigorous) (Skill 5.6)

    A.   they do not aid in recombination procedures

    B.   they are used in genetic engineering

    C.   they are named after the bacteria in which they naturally occur

    D.   they identify and splice certain base sequences on DNA

Answer: A. they do not aid in recombination procedures
A restriction enzyme is a bacterial enzyme that cuts foreign DNA at specific locations. The splicing of restriction fragments into a plasmid results in a recombinant plasmid.

76. A virus that can remain dormant until a certain environmental condition causes its rapid increase is said to be *(Average Rigor)* *(Skill 6.1)*

 A. lytic

 B. benign

 C. saprophytic

 D. lysogenic

Answer: D. lysogenic
Lysogenic viruses remain dormant until something initiates it to break out of the host cell.

77. Which is not considered to be a morphological type of bacteria? *(Average Rigor) (Skill 6.3)*

 A. obligate

 B. coccus

 C. spirillum

 D. bacillus

Answer: A. bacillus
Morphology is the shape of an organism. Obligate is a term used when describing dependence on something. Coccus is a round bacterium, spirillum is a spiral shaped bacterium, and bacillus is a rod shaped bacterium.

78. Antibiotics are effective in fighting bacterial infections due to their ability to *(Rigorous) (Skill 6.5)*

 A. interfere with DNA replication in the bacteria

 B. prevent the formation of new cell walls in the bacteria

 C. disrupt the ribosome of the bacteria

 D. All of the above.

Answer: D. all of the above
Antibiotics can destroy the bacterial cell wall, interfere with bacterial DNA replication, and disrupt the bacterial ribosome without affecting the host cells.

79. **Bacteria commonly reproduce by a process called binary fission. Which of the following best defines this process?** *(Average Rigor) (Skill 4.3)*

    A.   viral vectors carry DNA to new bacteria

    B.   DNA from one bacterium enters another

    C.   DNA doubles and the bacterial cell divides

    D.   DNA from dead cells is absorbed into bacteria

Answer: C. DNA doubles and the bacterial cell divides
Binary fission is the asexual process in which the bacteria divide in half after the DNA doubles. This results in an exact clone of the parent cell.

80. **All of the following are examples of a member of Kingdom Fungi except** *(Average Rigor) (Skill 4.6)*

    A.   mold

    B.   algae

    C.   mildew

    D.   mushrooms

Answer: B. algae
Mold, mildew, and mushrooms are all fungi. Brown algae and golden algae are members of the kingdom protista and green algae are members of the plant kingdom.

81. **Protists are classified into major groups according to** *(Average Rigor) (Skill 4.6)*

   A.   their method of obtaining nutrition

   B.   reproduction

   C.   metabolism

   D.   their form and function

Answer: A. their method of obtaining nutrition
The chaotic status of names and concepts of the higher classification of the protists reflects their great diversity in form, function, and lifestyles. The protists are often grouped as algae (plant-like), protozoa (animal-like), or fungus-like based on how they obtain nutrition and the similarity of their lifestyle and characteristics to these more derived groups.

82. **In comparison to protist cells, moneran cells**

   I. are usually smaller
   II. evolved later
   III. are more complex
   IV. contain more organelles
   (Rigorous) (Skill 6.1)

   A.   I

   B.   I and II

   C.   II and III

   D.   I and IV

Answer: A. I
Moneran cells are almost always smaller than protists. Moneran cells are prokaryotic; therefore they are less complex and have no organelles. Prokaryotes were the first cells on Earth and therefore evolved before than the eukaryotic protists.

83.   Spores characterize the reproduction mode for which of the following group of plants? *(Average Rigor) (Skill 7.9)*

A.   algae

B.   flowering plants

C.   conifers

D.   ferns

Answer: D. ferns
Ferns are non-seeded vascular plants. All plants in this group have spores and require water for reproduction. Algae, flowering plants, and conifers are not in this group of plants.

84.   Water movement to the top of a twenty foot tree is most likely due to which principle? *(Rigorous) (Skill 3.1)*

A.   osmostic pressure

B.   xylem pressure

C.   capillarity

D.   transpiration

Answer: D. transportation
Xylem is the tissue that transports water upward. Transpiration is the force that pulls the water upwards. Transpiration is the evaporation of water from leaves.

85.   What are the stages of development from the egg to the plant? *(Average Rigor) (Skill 7.9)*

A.   morphogenesis, growth, and cellular differentiation

B.   cell differentiation, growth, and morphogenesis

C.   growth, morphogenesis, and cellular differentiation

D.   growth, cellular differentiation, and morphogenesis

Answer: C. growth, morphogenesis, and cellular differentiation
The development of the egg to form a plant occurs in three stages: growth; morphogenesis, the development of form; and cellular differentiation, the acquisition of a cell's specific structure and function.

86. **In angiosperms, the food for the developing plant is found in which of the following structures?** *(Average Rigor) (Skill 7.10)*

 A. ovule

 B. endosperm

 C. male gametophyte

 D. cotyledon

Answer: B. endosperm
The endosperm is a product of double fertilization. It is the food supply for the developing plant.

87. **The process in which pollen grains are released from the anthers is called** *(Easy) (Skill 7.9)*

 A. pollination

 B. fertilization

 C. blooming

 D. dispersal

Answer: A. pollination
Pollen grains are released from the anthers during pollination and carried by animals and the wind to land on the carpels.

88. **Which of the following is not a characteristic of a monocot?** *(Average Rigor) (Skill 7.6)*

 A. parallel veins in leaves

 B. petals of flowers occur in multiples of 4 or 5

 C. one seed leaf

 D. vascular tissue scattered throughout the stem

Answer: B. petals of flowers occur in multiples of 4 or 5
Monocots have one cotelydon, parallel veins in their leaves, and their flower petals are in multiples of threes. Dicots have flower petals in multiples of fours and fives.

89. **What controls gas exchange on the bottom of a plant leaf?** *(Easy)* *(Skill 7.3)*

    A.   stomata

    B.   epidermis

    C.   collenchyma and schlerenchyma

    D.   palisade mesophyll

Answer: A. stomata
Stomata provide openings on the underside of leaves for oxygen to move in or out of the plant and for carbon dioxide to move in.

90. **How are angiosperms different from other groups of plants?** *(Average Rigor) (Skill 7.5)*

    A.   presence of flowers and fruits

    B.   production of spores for reproduction

    C.   true roots and stems

    D.   seed production

Answer: A. presence of flowers and fruits
Angiosperms do not have spores for reproduction. They do have true roots and stems as do all vascular plants. They do have seed production as do the gymnosperms. The presence of flowers and fruits is the difference between angiosperms and other plants.

91. **Generations of plants alternate between** *(Average Rigor) (Skill 7.7)*

    A.   angiosperms and bryophytes

    B.   flowering and non-flowering stages

    C.   seed bearing and spore bearing plants

    D.   haploid and diploid stages

Answer: D. haploid and diploid stages
Reproduction of plants is accomplished through alteration of generations. Simply stated, a haploid stage in the plant's life history alternates with a diploid stage.

92. **Double fertilization refers to which choice of the following?** *(Average Rigor) (Skill 7.9)*

    A.   two sperm fertilizing one egg

    B.   fertilization of a plant by gametes from two separate plants

    C.   two sperm enter the plant embryo sac; one sperm fertilizes the egg, the other forms the endosperm

    D.   the production of non-identical twins through fertilization of two separate eggs

Answer: C. two sperm enter the plant embryo sac; one sperm fertilizes the egg, the other forms the endosperm
In angiosperms, double fertilization is when an ovum is fertilized by two sperm. One sperm produces the new plant and the other forms the food supply for the developing plant (endosperm).

93. **Characteristics of coelomates include:**

    I. no true digestive system
    II. two germ layers
    III. true fluid filled cavity
    IV. three germ layers
    (Rigorous) (Skill 8.2)

    A.   I

    B.   II and IV

    C.   IV

    D.   III and IV

Answer: D. III and IV
Coelomates are triplobastic animals (3 germ layers). They have a true fluid filled body cavity called a coelom.

94.   Which phylum accounts for 85% of all animal species? *(Rigorous)*
       (Skill 8.2)

       A.   Nematoda

       B.   Chordata

       C.   Arthropoda

       D.   Cnidaria

Answer: C. Arthropoda
The arthropoda phylum consists of insects, crustaceans, and spiders. They are
the largest group in the animal kingdom.

95.   Which is the correct statement regarding the human nervous system
       and the human endocrine system? *(Rigorous) (Skill 8.6)*

       A.   the nervous system maintains homeostasis whereas the endocrine
             system does not

       B.   endocrine glands produce neurotransmitters whereas nerves
             produce hormones

       C.   nerve signals travel on neurons whereas hormones travel through
             the blood

       D.   he nervous system involves chemical transmission whereas the
             endocrine system does not

**Answer: C. nerve signals travel on neurons whereas hormones travel
through the blood**
In the human nervous system, neurons carry nerve signals to and from the cell
body. Endocrine glands produce hormones that are carried through the body in
the bloodstream.

96.   **A muscular adaptation to move food through the digestive system is called** *(Average Rigor) (Skill 8.10)*

A.   peristalsis

B.   passive transport

C.   voluntary action

D.   bulk transport

Answer: A. peristalsis
Peristalsis is a process of wave-like contractions. This process allows food to be carried down the pharynx and though the digestive tract.

97.   **The role of neurotransmitters in nerve action is** *(Rigorous) (Skill 8.6)*

A.   turn off sodium pump

B.   turn off calcium pump

C.   send impulse to neuron

D.   send impulse to the body

Answer: A. turn off sodium pump
The neurotransmitters turn off the sodium pump which results in depolarization of the membrane.

98.   **Fats are broken down by which substance?** *(Rigorous) (Skill 3.4)*

A.   bile produced in the gall bladder

B.   lipase produced in the gall bladder

C.   glucagons produced in the liver

D.   bile produced in the liver

Answer: D. bile produced in the liver
The liver produces bile which is stored in the gall bladder, then released into the small intestine where it breaks emulsifies (physically breaks down) fatty acids.

100. Fertilization in humans usually occurs in the *(Average Rigor) (Skill 4.4)*

    A.   uterus

    B.   ovary

    C.   fallopian tubes

    D.   vagina

Answer: C. fallopian tubes
Fertilization of the egg by the sperm normally occurs in the fallopian tube. The fertilized egg is then implanted on the uterine lining for development.

100. All of the following are found in the dermis layer of skin except *(Average Rigor) (Skill 8.7)*

    A.   sweat glands

    B.   keratin

    C.   hair follicles

    D.   blood vessels

Answer: B. keratin
Keratin is a water proofing protein found in the epidermis.

101. Which is the correct sequence of embryonic development in a frog? *(Rigorous) (Skill 1.17)*

    A.   cleavage – blastula – gastrula

    B.   cleavage – gastrula – blastula

    C.   blastula – cleavage – gastrula

    D.   gastrula – blastula – cleavage

Answer: A.   cleavage – blastula – gastrula
Animals go through several stages of development after fertilization of the egg cell. The first step is cleavage which continues until the egg becomes a blastula. The blastula is a hollow ball of undifferentiated cells. Gastrulation is the next step. This is the time of tissue differentiation into the separate germ layers: the endoderm, mesoderm, and ectoderm.

102. **Food is carried through the digestive tract by a series of wave-like contractions. This process is called** *(Average Rigor) (Skill 8.5)*

    A.   peristalsis

    B.   chyme

    C.   digestion

    D.   absorption

Answer: A. peristalsis
Peristalsis is the process of wave-like contractions that moves food through the digestive tract.

103. **Movement is possible by the action of muscles pulling on** *(Easy) (Skill 8.7)*

    A.   skin

    B.   bones

    C.   joints

    D.   ligaments

Answer: B. bones
The muscular system's function is for movement. Skeletal muscles are attached to bones and are responsible for their movement.

104. **All of the following are functions of the skin except** *(Easy) (Skill 8.9)*

    A.   storage

    B.   protection

    C.   sensation

    D.   regulation of temperature

Answer: A. storage
Skin is a protective barrier against infection. It contains hair follicles that respond to sensation and it plays a role in thermoregulation.

105. **Hormones are essential to the regulation of reproduction. What organ is responsible for the release of hormones for sexual maturity?** *(Rigorous) (Skill 8.10)*

    A.   pituitary gland

    B.   hypothalamus

    C.   pancreas

    D.   thyroid gland

Answer: B. hypothalamus
The hypothalamus begins secreting hormones that help mature the reproductive system and development of the secondary sex characteristics.

106. **A bicyclist has a heart rate of 110 beats per minute and a stroke volume of 85 mL per beat. What is the cardiac output?** *(Rigorous) (Skill 8.4)*

    A.   9.35 L/min

    B.   1.29 L/min

    C.   0.772 L/min

    D.   129 L/min

Answer: A. 9.35 L/min
The cardiac output is the volume of blood per minute that is pumped into the systemic circuit. This is determined by the heart rate and the stroke volume. Multiply the heart rate by the stroke volume. 110 * 85 = 9350 mL/min. Divide by 1000 to get units of liters. 9350/1000 = 9.35 L/min.

107. After sea turtles are hatched on the beach, they start the journey to the ocean. This is due to *(Easy) (Skill 8.12)*

A.   innate behavior

B.   territoriality

C.   the tide

D.   learned behavior

Answer: A. innate behavior
Innate behavior is inborn or instinctual. The baby sea turtles did not learn from their mother. They immediately knew to head towards the ocean once they hatched.

108. A school age boy had the chicken pox as a baby. He will most likely not get this disease again because of *(Easy) (Skill 3.10)*

A.   passive immunity

B.   vaccination

C.   antibiotics

D.   active immunity

Answer: D. active immunity
Active immunity develops after recovery from an infectious disease, such as the chicken pox, or after vaccination. Passive immunity may be passed from one individual to another (from mother to nursing child).

109. **High humidity and temperature stability are present in which of the following biomes?** *(Average Rigor) (Skill 9.1)*

    A.   taiga

    B.   deciduous forest

    C.   desert

    D.   tropical rain forest

Answer: D. tropical rain forest
A tropical rain forest is located near the equator. Its temperature is at a constant 25 degrees C and the humidity is high due to the rainfall that exceeds 200 cm per year.

110. **The biological species concept applies to** *(Average Rigor) (Skill 10.2)*

    A.   asexual organisms

    B.   extinct organisms

    C.   sexual organisms

    D.   fossil organisms

Answer: C. sexual organisms
The biological species concept states that a species is a reproductive community of populations that occupy a specific niche in nature. It focuses on reproductive isolation of populations as the primary criterion for recognition of species status. The biological species concept does not apply to organisms that are completely asexual in their reproduction, fossil organisms, or distinctive populations that hybridize.

111. **Which term is not associated with the water cycle?** *(Average Rigor)*
   (Skill 9.3)

   A.   precipitation

   B.   transpiration

   C.   fixation

   D.   evaporation

Answer: C. fixation
Water is recycled through the processes of evaporation and precipitation.
Transpiration is the evaporation of water from leaves. Fixation is not associated
with the water cycle.

112. **All of the following are density independent factors that affect a
   population except** *(Average Rigor) (Skill 9.5)*

   A.   temperature

   B.   rainfall

   C.   predation

   D.   soil nutrients

Answer: C. predation
As a population increases, the competition for resources is intense and the
growth rate declines. This is a density-dependent factor. An example of this
would be predation. Density-independent factors affect the population regardless
of its size. Examples of density-independent factors are rainfall, temperature,
and soil nutrients.

113. In the growth of a population, the increase is exponential until carrying capacity is reached. This is represented by a (n) *(Average Rigor) (Skill 9.5)*

    A.   S curve

    B.   J curve

    C.   M curve

    D.   L curve

Answer: A. S curve
An exponentially growing population starts off with little change and then rapidly increases. The graphic representation of this growth curve has the appearance of a "J". However, as the carrying capacity of the exponentially growing population is reached, the growth rate begins to slow down and level off. The graphic representation of this growth curve has the appearance of an "S".

114. Primary succession occurs after *(Average Rigor) (Skill 9.7)*

    A.   nutrient enrichment

    B.   a forest fire

    C.   bare rock is exposed after a water table recedes

    D.   a housing development is built

Answer: C. bare rock is exposed after a wave table recedes
Primary succession occurs where life never existed before, such as flooded areas or a new volcanic island. It is only after the water recedes that the rock is able to support new life.

115. **Crabgrass – grasshopper – frog – snake – eagle; If DDT were present in an ecosystem, which organism would have the highest concentration in its system?** *(Rigorous) (Skill 9.13)*

    A.   grasshopper

    B.   eagle

    C.   frog

    D.   crabgrass

Answer: B. eagle
Chemicals and pesticides accumulate along the food chain. Tertiary consumers have more accumulated toxins than animals at the bottom of the food chain.

116. **Which trophic level has the highest ecological efficiency?** *(Average Rigor) (Skill 9.4)*

    A.   decomposers

    B.   producers

    C.   tertiary consumers

    D.   secondary consumers

Answer: B. producers
The amount of energy that is transferred between trophic levels is called the ecological efficiency. The visual of this is represented in a pyramid of productivity. The producers have the greatest amount of energy and are at the bottom of this pyramid.

117.   A clownfish is protected by the sea anemone's tentacles. In turn, the anemone receives uneaten food from the clownfish. This is an example of (*Easy*) (*Skill 9.6*)

   A.   mutualism

   B.   parasitism

   C.   commensalisms

   D.   competition

Answer: A. mutualism
Neither the clownfish nor the anemone cause harmful effects towards one another and they both benefit from their relationship. Mutualism is when two species that occupy a similar space benefit from their relationship.

118.   If the niches of two species overlap, what usually results? (*Average Rigor*) (*Skill 7.4*)

   A.   a symbiotic relationship

   B.   cooperation

   C.   competition

   D.   a new species

Answer: C. competition
Two species that occupy the same habitat or eat the same food are said to be in competition with each other.

119. Oxygen created in photosynthesis comes from the breakdown of *(Rigorous) (Skill 3.6)*

   A.   carbon dioxide

   B.   water

   C.   glucose

   D.   carbon monoxide

Answer: B. water
In photosynthesis, water is split; the hydrogen atoms are pulled to carbon dioxide which is taken in by the plant and ultimately reduced to make glucose. The oxygen from the water is given off as a waste product.

120. Which photosystem makes ATP? *(Average Rigor) (Skill 3.6)*

   A.   photosystem I

   B.   photosystem II

   C.   photosystem III

   D.   photosystem IV

Answer: A. photosystem I
Photosystem I is composed of a pair of chlorophyll *a* molecules. It makes ATP whose energy is needed to build glucose.

121. All of the following gasses made up the primitive atmosphere except *(Average Rigor) (Skill 3.13)*

   A.   ammonia

   B.   methane

   C.   oxygen

   D.   hydrogen

swer: C. oxygen
he 1920s, Oparin and Haldane were to first to theorize that the primitive
osphere was a reducing atmosphere with no oxygen. The gases were rich in
drogen, methane, water, and ammonia.

122. **The Endosymbiotic theory states that** *(Average Rigor) (Skill 4.3)*

   A.   eukaryotes arose from prokaryotes

   B.   animals evolved in close relationships with one another

   C.   the prokaryotes arose from eukaryotes

   D.   life arose from inorganic compounds

Answer: A. eukaryotes arose from prokaryotes
The Endosymbiotic theory of the origin of eukaryotes states that eukaryotes arose from symbiotic groups of prokaryotic cells. According to this theory, smaller prokaryotes lived within larger prokaryotic cells, eventually evolving into chloroplasts and mitochondria.

123. **Which aspect of science does not support evolution?** *(Easy) (Skill 1.17)*

   A.   comparative anatomy

   B.   organic chemistry

   C.   comparison of DNA among organisms

   D.   analogous structures

Answer: B. organic chemistry
Comparative anatomy is the comparison of characteristics of the anatomies of different species. This includes homologous structures and analogous structures. The comparison of DNA between species is the best known way to place species on the evolutionary tree. Organic chemistry has nothing to do with evolution.

124. **Evolution occurs in** *(Easy) (Skill 1.17)*

 A. individuals

 B. populations

 C. organ systems

 D. cells

Answer: B. populations
Evolution is a change in genotype over time. Gene frequencies shift and change from generation to generation. Populations evolve, not individuals.

125. **Which process contributes to the large variety of living things in the world today?** *(Average Rigor) (Skill 1.17)*

 A. meiosis

 B. asexual reproduction

 C. mitosis

 D. alternation of generations

Answer: A. meiosis
During meiosis prophase I crossing over occurs. This exchange of genetic material between homologues increases diversity.

126. **The wing of bird, human arm and whale flipper have the same bone structure. These are called** *(Easy) (Skill 1.17)*

 A. polymorphic structures

 B. homologous structures

 C. vestigial structures

 D. analogous structures

·swer: B. homologous structures
 ·ologous characteristics have the same genetic basis (leading to similar
 arances) but are used for a different function.

127. **Which biome is the most prevalent on Earth?** *(Easy) (Skill 9.1)*

   A.   marine

   B.   desert

   C.   savanna

   D.   tundra

Answer: A.   marine
The marine biome covers 75% of the Earth. This biome is organized by the depth of water.

128. **Which of the following is not an abiotic factor?** *(Average Rigor) (Skill 9.7)*

   A.   temperature

   B.   rainfall

   C.   soil quality

   D.   bacteria

Answer: D. bacteria
Abiotic factors are non-living aspects of an ecosystem. Bacteria is an example of a biotic factor—a living thing in an ecosystem.

129. **DNA synthesis results in a strand that is synthesized continuously. This is the** *(Rigorous) (Skill 5.2)*

   A.   lagging strand

   B.   leading strand

   C.   template strand

   D.   complementary strand

Answer: B. leading strand
As DNA synthesis proceeds along the replication fork, one strand is replicated continuously (the leading strand) and the other strand is replicated discontinuously (lagging strand).

130. **Using a gram staining technique, it is observed that E. coli stains pink. It is therefore** *(Easy) (Skill 6.1)*

   A.   gram positive

   B.   dead

   C.   gram negative

   D.   gram neutral

Answer: C. gram negative
A Gram positive bacterium absorbs the stain and appears purple under a microscope because of its cell wall made of peptidoglycan. A Gram negative bacterium does not absorb the stain because of its more complex cell wall. These bacteria appear pink under a microscope.

131. **A light microscope has an ocular of 10X and an objective of 40X. What is the total magnification?** *(Easy) (Skill 1.1)*

   A.   400X

   B.   30X

   C.   50X

   D.   4000X

Answer: A. 400X
To determine the total magnification of a microscope, multiply the ocular lens by the objective lens. Here, the ocular lens is 10X and the objective lens is 40X.

   (10X) X (40X) = 400X total magnification

132. Three plants were grown. The following data was taken. Determine the mean growth.

Plant 1: 10cm   Plant 2: 20cm   Plant 3: 15cm

(Rigorous) (Skill 1.4)

A.   5 cm

B.   45 cm

C.   12 cm

D.   15 cm

Answer: D. 15 cm
The mean growth is the average of the three growth heights.

$$\frac{10 + 20 + 15}{3} = 15\text{cm average height}$$

133. Electrophoresis separates DNA on the basis of *(Average Rigor) (Skill 1.5)*

A.   amount of current

B.   molecular size

C.   positive charge of the molecule

D.   solubility of the gel

Answer: B. molecular size
Electrophoresis uses electrical charges of molecules to separate them according to their size.

134. The reading of a meniscus in a graduated cylinder is done at the
(Easy) (Skill 1.3)

A. top of the meniscus

B. middle of the meniscus

C. bottom of the meniscus

D. closest whole number

Answer: C. bottom of the meniscus
The graduated cylinder is the common instrument used for measuring volume. It is important for the accuracy of the measurement to read the volume level of the liquid at the bottom of the meniscus. The meniscus is the curved surface of the liquid.

135. Two hundred plants were grown. Fifty plants died. What percentage of the plants survived? *(Easy) (Skill 1.4)*

A. 40%

B. 25%

C. 75%

D. 50%

Answer: C. 75%
This is a proportion. If 50 plants died, then 200 – 50 = 150 survived. The number of survivors is the numerator and the total number of plants grown is the denominator.

$$\frac{150}{200} = 0.75 \text{ Multiply by 100 to get percent} = 75\% \text{ survive}$$

136. **Which is not a correct statement regarding the use of a light microscope?** *(Easy) (Skill 1.1)*

    A.   carry the microscope with two hands

    B.   store on the low power objective

    C.   clean all lenses with lens paper

    D.   Focus first on high power

Answer: D. focus first on high power
Always begin focusing on low power. This allows for the observation of microorganisms in a larger field of view. Switch to high power once you have a microorganism in view on low power.

137. **Spectrophotometry utilizes the principle of** *(Easy) (Skill 1.5)*

    A.   light transmission

    B.   molecular weight

    C.   solubility of the substance

    D.   electrical charges

Answer: A. light transmission
Spectrophotometry uses percent of light at different wavelengths absorbed and transmitted by a pigment solution.

138. **Paper chromotography is most often associated with the separation of** *(Average Rigor) (Skill 1.5)*

    A.   nutritional elements

    B.   DNA

    C.   proteins

    D.   plant pigments

Answer: D. plant pigments
Paper chromatography uses the principles of capillarity to separate substan such as plant pigments. Molecules of a larger size will move slower up the paper, whereas smaller molecules will move more quickly producing lines o pigment.

139. **A genetic engineering advancement in the medical field is** *(Easy)* *(Skill 5.6)*

    A.   gene therapy

    B.   pesticides

    C.   degradation of harmful chemicals

    D.   antibiotics

Answer: A. gene therapy
Gene therapy is the introduction of a normal allele to the somatic cells to replace a defective allele. The medical field has had success in treating patients with a single enzyme deficiency disease. Gene therapy has allowed doctors and scientists to introduce a normal allele that would provide the missing enzyme.

140. **Which scientists are credited with the discovery of the structure of DNA?** *(Easy) (Skill 1.14)*

    A.   Hershey & Chase

    B.   Sutton & Morgan

    C.   Watson & Crick

    D.   Miller & Fox

Answer: C. Watson & Crick
In the 1950s, James Watson and Francis Crick discovered the structure of a DNA molecule as that of a double helix.

141. **Negatively charged particles that circle the nucleus of an atom are called** *(Easy) (Skill 1.3)*

    A. neutrons

    B. neutrinos

    C. electrons

    D. protons

Answer: C. electrons
Neutrons and protons make up the core of an atom. Neutrons have no charge and protons are positively charged. Electrons are the negatively charged particles around the nucleus.

142. **The shape of a cell depends on its** *(Easy) (Skill 4.2)*

    A. function

    B. structure

    C. age

    D. size

Answer: A. functions
In most living organisms, its structure is based on its function.

143. **The most ATP is generated through** *(Average Rigor) (Skill 3.4)*

    A. fermentation

    B. glycolysis

    C. chemiosmosis

    D. Krebs cycle

Answer: C. chemiosmosis
The electron transport chain uses electrons to pump hydrogen ions across the mitochondrial membrane. This ion gradient is used to form ATP in a process called chemiosmosis. ATP is generated by the movement of hydrogen ions NADH and $FADH_2$. This yields 34 ATP molecules.

144. **In DNA, adenine bonds with ____, while cytosine bonds with ____.** *(Average Rigor) (Skill 5.2)*

   A.   thymine/guanine

   B.   adenine/cytosine

   C.   cytosine/adenine

   D.   guanine/thymine

Answer: A. thymine/guanine
In DNA, adenine pairs with thymine and cytosine pairs with guanine because of their nitrogenous base structures.

145. **The individual parts of cells are best studied using a (n) *(Easy) (Skill 1.2)***

   A.   ultracentrifuge

   B.   phase-contrast microscope

   C.   CAT scan

   D.   electron microscope

Answer: D. electron miscroscope
The scanning electron microscope uses a beam of electrons to pass through the specimen. The resolution is about 1000 times greater than that of a light microscope. This allows the scientist to view extremely small objects, such as the individual parts of a cell.

146. **Thermoacidophiles are *(Rigorous) (Skill 6.1)***

   A.   prokaryotes

   B.   eukaryotes

   C.   protists

   D.   archaea

r: D. archaea
ɔacidophiles, methanogens, and halobacteria are members of the archaea
   They are as diverse from prokaryotes as prokaryotes are to eukaryotes.

147. **Which of the following is not a type of fiber that makes up the cytoskeleton?** *(Rigorous) (Skill 4.2)*

    A.   vacuoles

    B.   microfilaments

    C.   microtubules

    D.   intermediate filaments

Answer: A. vacuoles
Vacuoles are mostly found in plants and hold stored food and pigments. The other three choices are fibers that make up the cytoskeleton found in both plant and animal cells.

148.   **Viruses are made of** *(Easy) (Skill 5.7)*

    A.   a protein coat surrounding a nucleic acid

    B.   DNA, RNA and a cell wall

    C.   a nucleic acid surrounding a protein coat

    D.   protein surrounded by DNA

Answer: A.   a protein coat surrounding a nucleic acid
Viruses are composed of a protein coat and a nucleic acid; either RNA or DNA.

149.   **Reproductive isolation results in** *(Easy) (Skill 10.2)*

    A.   extinction

    B.   migration

    C.   follilization

    D.   speciation

Answer: D. speciation
Reproductive isolation is caused by any factor that impedes two species from producing viable, fertile hybrids. Reproductive isolation of populations is the primary criterion for recognition of species status.

150. **This protein structure consists of the coils and folds of polypeptide chains. Which is it?** *(Average Rigor) (Skill 3.1)*

    A.   secondary structure

    B.   quaternary structure

    C.   tertiary structure

    D.   primary structure

Answer: A. secondary structure
Primary structure is the protein's unique sequence of amino acids. Secondary structure is the coils and folds of polypeptide chains. The coils and folds are the result of hydrogen bonds along the polypeptide backbone. Tertiary structure is formed by bonding between the side chains of the amino acids. Quaternary structure is the overall structure of the protein from the aggregation of two or more polypeptide chains.

CPSIA information can be obtained
at www.ICGtesting.com
Printed in the USA
BVHW091204190719

553928BV00014B/906/P